Development Periods and Genetic Mechanisms of Fractures in Volcanic Reservoirs in the Hala'alate Mountain Area, Northwestern Margin of the Junggar Basin

Haifeng Yuan

SCIENCE PRESS

Beijing

Responsible Editor: Qiao Huang

Copyright © 2016 by Science Press
Published by Science Press
16 Donghuangchenggen North Street
Beijing 100717, P.R. China

Printed in Chengdu

All rights reserved. No part of this publication may be reproduced, stored in a retrieval system, or transmitted in any form or by any means, electronic, mechanical, photocopying, recording or otherwise, without the prior written permission of the copyright owner.

ISBN 978-7-03-049577-8

Preface

The oil and gas exploration in the Hala'alate Mountain structural belt, northwestern margin of the Junggar Basin began in the early 1950s. This structural belt has abundant oil and gas resources and huge exploration potential. The overall structural framework in the Hala'alate Mountain area is currently a large superimposed nappe structure mainly composed of superimposed multiphase thrust nappes. The Hala'alate Mountain structure has experienced superimposed reformation due to multiphase tectonic movements in Hercynian, Indosinian, Yanshan and Himalayan Period, as a result, complex fault systems and associated fracture systems have been formed. As fractures are important oil and gas migration pathways and reservoir space, the genetic mechanisms and development periods of fractures and their relationship with oil and gas accumulation are key to oil and gas exploration. Therefore, the following issues have been investigated in the book in the hope to provide reference and guidance for oil and gas exploration in the Hala'alate Mountain area.

(1) Rock types, and pore types in the Hala'alate Mountain area and the controlling factors on their development have been sorted out, and fracture occurrence and filling situation have been examined by using core observation and imaging logging data.

(2) The geochemical characteristics of the main source rocks in the Hala'alate Mountain structure have been analyzed, the main oil and gas sources in this area have been figured out based on the characteristics and differences of biomarkers, the carrier systems and migration pathways of oil and gas have been sought out, Sr, C and O isotopes of fracture fillings and characteristics of fluid inclusions have been analyzed, and fluid source and filling period of filling minerals have been investigated prominently.

(3) The pattern and evolution process of structures in the Hala'alate Mountain area have been analyzed, based on the mechanical differences of different lithologies and the imaging logging data, the development sequence of main faults and their controlling effects on fracture development have been investigated, and the relationship between fracture development periods and hydrocarbon charging has been figured out.

We would like to thank the experts of Oil and Gas Exploration Management Center,

Sinopec Shengli Oilfield Company for providing basic data and precious advice in the process of project research.

Limited by the authors' ability, there could be some unintended mistakes in the book, and your comments and suggestions are gratefully accepted.

Haifeng Yuan
March, 2016 in Chengdu

Contents

Chapter 1　Regional Geological Setting ··1

1.1　Geological setting of the Junggar Basin ··1
　　1.1.1　Overview of the basin···1
　　1.1.2　Structural evolution and structural unit division ···2
1.2　Geological setting of the northwestern margin of the Junggar Basin ···································4
　　1.2.1　Geological setting···4
　　1.2.2　Structural evolution ··6

Chapter 2　Reservoir Petrological Characteristics and Reservoir Space Types ················10

2.1　Reservoir petrology··10
　　2.1.1　Volcanic breccia ···10
　　2.1.2　Andesite ···13
　　2.1.3　Basalt ··15
　　2.1.4　Tuff ···16
　　2.1.5　Clastic rock ···17
2.2　Reservoir space types··19
　　2.2.1　Primary reservoir space ··20
　　2.2.2　Secondary reservoir space ··23
2.3　Factors influencing reservoir space development ··27
　　2.3.1　Effect of lithology and lithofacies··27
　　2.3.2　Effect of dissolution ··31
　　2.3.3　Effect of tectonic activity ···32

Chapter 3　Oil-source Correlation and Migration Pathway Tracing ································34

3.1　Geochemistry of source rocks ···35
　　3.1.1　Evaluation criterion of source rocks ··35
　　3.1.2　Evaluation of source rocks ··36
　　3.1.3　Biomarkers of source rocks ··42
3.2　Geochemical features of crude oil and oil-source correlation ···45
　　3.2.1　Physical properties of crude oil ···45

  3.2.2 Geochemical features of crude oil and oil-source analysis ················· 45
3.3 Migration pathways and tracing of crude oil ································· 51
  3.3.1 Conducting system of crude oil ··· 51
  3.3.2 Tracing of crude oil migration pathways ······························· 55

Chapter 4 Development Features of Reservoir Fractures ···················· 59

4.1 Development features of core fractures ·· 59
  4.1.1 Identification and statistic of fractures in coring sections ··············· 60
  4.1.2 Filling features of fractures in coring sections ···························· 61
4.2 Imaging log responses of fractures ··· 62
  4.2.1 Imaging logging response of typical drilling fractures ···················· 65
  4.2.2 High-conductivity fractures and high-resistivity fractures ··············· 77

Chapter 5 Geochemistry of Fracture Fillings ································· 87

5.1 Isotope geochemistry ·· 87
  5.1.1 Sr isotope ··· 87
  5.1.2 C and O isotopes ··· 93
5.2 Fluid inclusion geochemistry ·· 99
  5.2.1 Inclusion petrography ·· 100
  5.2.2 Fluorescence characteristics ··· 103
  5.2.3 Homogenization temperature ·· 105

Chapter 6 Mechanical Properties of Reservoir Rocks ······················ 107

6.1 Rock mechanics characteristics ·· 108
  6.1.1 Sample information and test ··· 108
  6.1.2 Sample test procedure and results ······································· 110
  6.1.3 Rock deformation characteristics ·· 112
  6.1.4 Rock mechanics parameters ··· 117
6.2 Factors affecting mechanical properties of rock ····························· 119
  6.2.1 Influence of rock types on mechanical properties ······················ 120
  6.2.2 Influence of pre-existing weak planes on mechanical properties of rock ········· 121
  6.2.3 Influence of confining pressure on mechanical properties of rock ······ 122

Chapter 7 Factors Affecting Fracture Development and Effectiveness ········ 124

7.1 Factors affecting fracture development ······································· 124
  7.1.1 Influence of lithology ··· 124
  7.1.2 Influence of formation thickness ·· 126
  7.1.3 Influence of structural position ·· 126

7.2 Development stages of fractures ··128
　　7.2.1 Sequence of fracture development··128
　　7.2.2 Tectonic evolution and fracture development stage ································141
7.3 Fracture effectiveness and distribution ···144
　　7.3.1 Analysis on fracture effectiveness ··144
　　7.3.2 Distribution of effective fractures···147
References ···151

Chapter 1
Regional Geological Setting

1.1 Geological setting of the Junggar Basin

1.1.1 Overview of the basin

Located at the west section of the large latitudinal oil rich "Gold Belt" in China, in Central Asia hinterland, the Junggar Basin, in a triangle shape, is held by Tianshan Mountains and Altai Mountains, surrounded by folded mountain systems. As the main part of Central Asia orogenic belt, it has experienced multiple phases of complex structural evolution in Hercynian, Indosinian, Yanshan and Himalayan Period. 700 km long from east to west, 370 km wide from south to north, with an area of $13 \times 10^4 \, km^2$, it is a large superimposed basin developed in Late Palaeozoic-Mesozoic

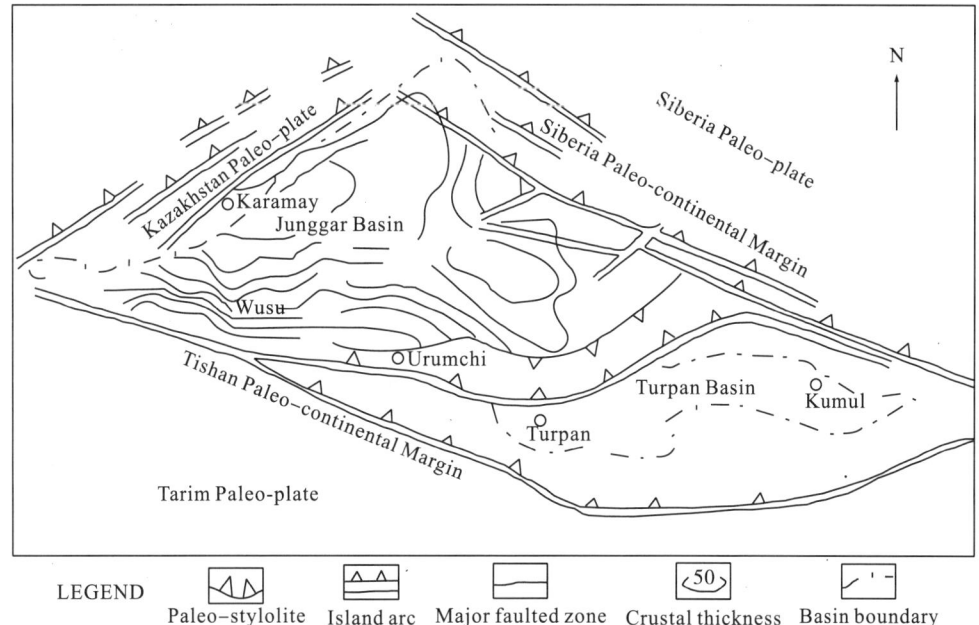

Figure 1-1 Tectonic setting of the Junggar Basin.

and Cenozoic. Surrounded by Palaeozoic folded mountain systems, it is one of the large complex superimposed hydrocarbon bearing basins in western China (Li et al., 2009). In the shape of a wedge, with basement of Pre-Cambrian crystalline rock and Early and Middle Palaeozoic fold systems, it is separated from Altai block of Siberia plate by the Irtysh-Zaysan Palaeozoic discordogenic fault and stylolite in the north, and separated from Tarim plate by the Palaeozoic discordogenic fault and stylolite of the northern margin of south Tianshan Mountains-Xingxing valley in the south. Tectonically, it is in the range of Kazakhstan plate (Chen et al., 2005; Chen and Wang, 2004) (Figure 1-1).

1.1.2 Structural evolution and structural unit division

At the early stage of Early Palaeozoic, a paleo-ocean occurred at the mountain along western boundary of current Junggar Basin, separating Junggar-Turpan microplate from Kazakhstan plate, and giving rise to the independent Junggar-Turpan microplate (Wang et al., 2013). At the end of Early Carboniferous, Junggar-Turpan microplate and Kazakhstan plate collided and connected together again, ushering a new development stage of western Junggar Basin (the collision marker in the stylolite, ophiolite suite is dated at C_1–C_2). During C_1–C_2, Siberia plate and Junggar plate converged and collided, kicking off the development of Kelameili nappe and eastern Juggar foreland system. During C_2–P_1, Junggar-Turpan plate and Tarim plate collided, ushering in the development of north Tianshan nappe structure and southern foreland basin system. Meanwhile, Bogda intercontinental rift was developed, inverted and closed due to plate subduction and collision, and consequently Bogda structural zone was formed, cutting Junggar-Turpan microplate into Junggar massif and Turpan massif, so afterwards, the basins in these two massifs moved into new and independent evolution period.

Since the Permian, the basin was transformed into a foreland basin with the gradual recession of sea water to the southeast, and eventually evolved into a large intracontinental depression lake basin. It was a foreland basin during the Early-Middle Permian and then was evolved into an intracontinental depression during the Late Permian-Triassic. During the Early Jurassic-Paleogene, the Junggar Basin entered into rift-depression basin cycle due to continental reconstruction, during which, it was characterized by extensional rifting at the early and middle stage of Jurassic, and afterwards, right lateral wrench movement happened and the extension was converted into compression gradually, only Changji sag had been in constant subsidence. After the Cretaceous, a united large intracontinental depression basin was formed, which, with thicker sediments, and center of subsidence moving to the south, had long maintained the structural framework of south depression-north slope. During the Neogene-Quaternary, with the large-scale uplifting of north Tianshan Mountains, the southern margin of the Junggar Basin experienced strong subsidence, multiple thrust fold belts were developed, and thick molasse formation was deposited in the piedmont sag.

Chapter 1 Regional Geological Setting

The Junggar Basin, formed in the compressional setting, is a large oil and gas bearing basin with the characteristics of complex superimposition. It is structurally characterized by multidirectional structure interlacing, multiple structural systems, multiple structural settings, multiphase structural evolution and multisource dynamics. From the Late Palaeozoic to the Quaternary, the Junggar Basin had experienced the tectonic movements of Hercynian, Indosinian, Yanshan and Himalayan, in which the late Hercynian Period was the key period for the formation and evolution of depression-uplift structural framework in this basin; and the basin suffered further superimposition and reworking (obvious in Zhundong) as a result of Indosinian-Yanshan movement. Himalayan movement had stronger action on the southern margin of the basin, but weaker effect on the other areas. The multicycle development of structures in the basin results in various structural assemblages and sedimentary systems, and has strictly controlled oil and gas accumulation and distribution.

There are multiple structural unit division programs for the Junggar Basin. According to structural evolution features of formations above Carboniferous, structure continuity, formation deposition, control degree of regional discordigenic faults to formation deposition etc, Yang Haibo (Yang et al., 2004) divided the structural units in the Junggar Basin into 6 first-order structural units, including two depressions (i.e., Wulungu depression and central depression), three uplifts (Luliang uplift, western uplift and eastern uplift) and one piedmont thrust belt (north Tianshan foreland thrust belt), and 44 second-order structural units (Figure 1-2).

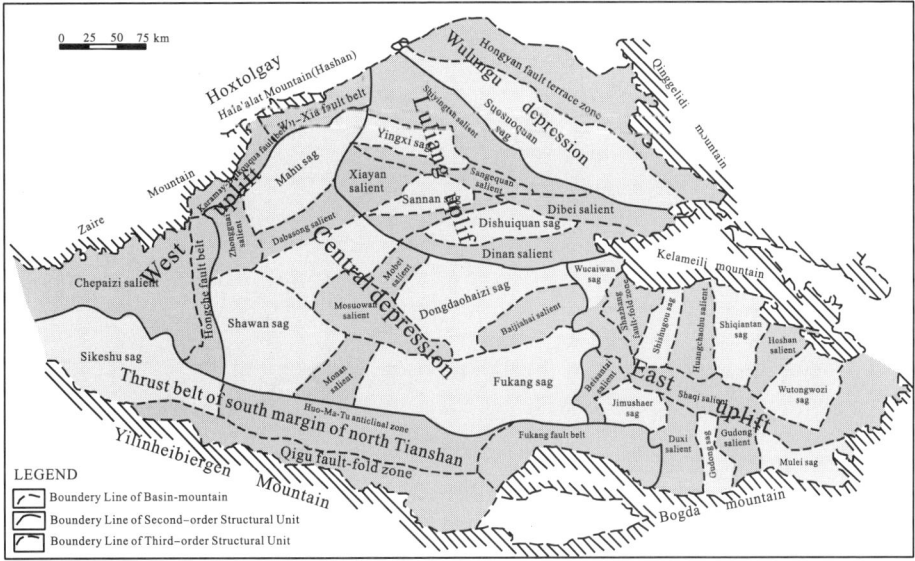

Figure 1-2 Structural unit division of the Junggar Basin (Yang et al., 2004).

1.2 Geological setting of the northwestern margin of the Junggar Basin

1.2.1 Geological setting

Bounded by the Hala'alate Mountain and Zaire Mountain, the northwestern margin of the Junggar Basin stretches from Kuitun in the southwest to Xiazijie in the northeast, in a NE band between west Junggar fold mountain system and Junggar block. From south to north, it is divided into 3 first-order structural units, i.e., Chepaizi salient (Hongche faulted zone included) in the southern section, Karamay-Baikouquan structural belt in the central section and Wuerhe-Xiazijie structural belt (called the Hala'alate Mountain structural belt in this book) in the northern section (Figure 1-3).

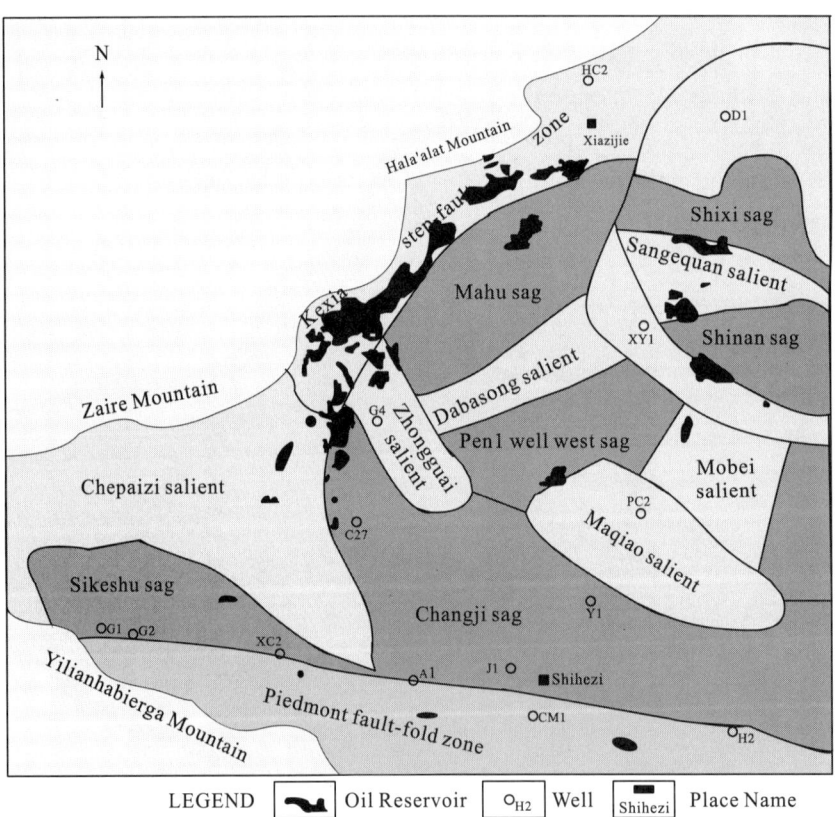

Figure 1-3   Structural location of the northwestern margin of the Junggar Basin (Sui, 2013).

The northwestern margin of the Junggar Basin is one of the important oil and gas exploration areas in western China. In the 1950s, Karamay Oilfield was discovered there. Later on, Chepaizi, Baijiantan, Baikouquan, Wuerhe, Fengcheng, Xiazijie, Mabei and Xiaoguai oilfields etc, have been discovered successively, indicating that the northwestern margin is one of the most enriched oil and gas areas in the Junggar Basin, and the area has abundant oil and gas resources and huge exploration potential.

In 2000, before Sinopec got involved in the exploration of this area, Xinjiang Oilfield had drilled 19 exploration wells in the Hala'alate Mountain structural belt, detecting oil and gas shows in the Jurassic and Cretaceous strata, but having no major discovery. Four wells (Well T1, H1, H2 and C4) were drilled by Xinjiang Oilfield Company in Chepaizi area, which had discovered oil and gas shows, but no commercial oil flow. During the 1980s-1990s, Chepaizi, Xiaoguai and Hongshanzui Oilfields were discovered successively. After 2000, comprehensive study and exploration had been carried out to figure out the source rock conditions, oil and gas migration conditions and trap types in this area, as a result, major exploration breakthroughs had been made in Chepaizi and the Hala'alate Mountain structural belts, such as the discovery of Chunguang, Chunfeng and Chunhui Oilfields, unveiling two exploration prospective areas with one hundred million tons of reserves (Sui, 2013).

The foreland thrust belt in the northwestern margin of the Junggar Basin is a large imbricate thrust system developing since the Late Carboniferous, and the thrust belt of the Hala'alat Mountain is the eastern section of the frontal structural belt of the foreland thrust belt. It is composed of the monoclinal zone of Ximisitai Mountain, the Heshituoluogai Basin, the imbricate structural zone of the Hala'alate Mountain and the west slope of Mahu sag (Figure 1-4). It is divided into two tectonic deformation layers (i.e., upper and lower tectonic deformation layers) with the detachment surface inside the Palaeozoic as the interface. In the upper structural deformation layer, there develop 3-4 rows of imbricate thrust structures, and all thrust faults converge at the detachment surface inside the Palaeozoic; the lower structural deformation layer may be a duplex where more than a dozen of Palaeozoic faulted blocks are superimposed. Its formation is caused by the subduction, consumption and collision of Junggar-Turpan plate ocean crust to Kazakhstan plate during the Late Palaeozoic, and thus the generation of collision uplift zones in this area and collision foreland sedimentary depression adjacent to the uplift zones, i.e., large overthrust structures in Chepaizi-Hongshanzui-Xiazijie area and Mahu sag (Lei et al., 2005). Since the Cenozoic, the northwestern margin of the Junggar Basin has shown the nature of foreland basin, where, with regional tectonic movements, a series of overthrust nappes and thrust faults have been formed.

Located at the northeastern section of the foreland thrust belt in the northwestern margin of the Junggar Basin, the Hala'alate Mountain, held between the Junggar Basin and the Heshituoluogai Basin, adjacent to Mahu sag to the south, is a typical buried faulted system with the nature of overthrust nappe. It is against Ke-Bai faulted fold belt to the west, bounded by Xiazijie faulted fold belt to the east and separated from the Heshituoluogai Basin to the north by Daerbute discordogenic fault (Liu, 2012), with trend turning from NE to nearly EW. With a total length of 70 km and width of 30 km, it covers an area of about 2000 km^2 (Figure 1-4).

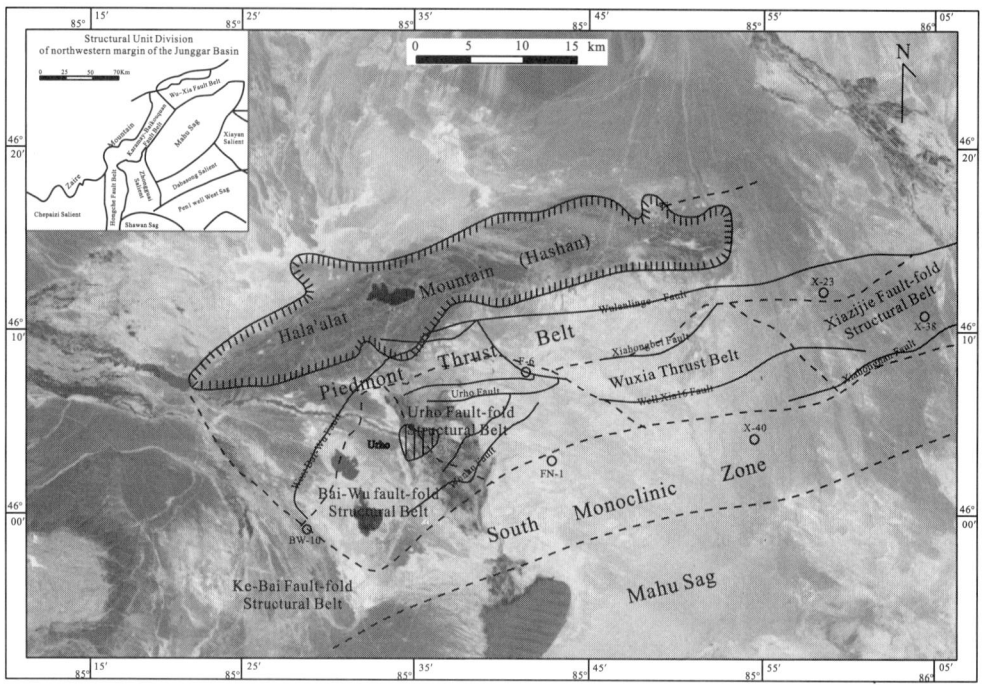

Figure 1-4 Geodetic position and tectonic zonation of Wu-Xia area in the northwestern margin of the Junggar Basin (Liu, 2012).

1.2.2 Structural evolution

The Hala'alate Mountain is a major structural unit in the Junggar Basin, where the formations are mainly Palaeozoic due to structural compression, arching and uplifting; and under the effect of long-term depositional hiatus and denudation, the outcropped strata are mostly Carboniferous (Liu, 2012), and the Upper Permian is seen sparsely in local parts. The outcropped strata are dominantly the Middle and Upper Carboniferous in the western section of the Hala'alate Mountain, the Middle Carboniferous in the central section and the Lower and Middle Carboniferous in the eastern section. The Upper Carboniferous strata are mainly composed of greyish green and brownish red sandstone and conglomerate. The Middle Carboniferous strata are mainly composed of greyish brown feldspathic clastic sandstone, mudstone with interbedded siltstone and andesite porphyrite in the upper part, dark grey and greyish green tuffaceous sandstone with interbedded shale in the middle part and grey and greyish green sandstone, siltstone and silty mudstone in the lower part. Differing widely in the west and in the east, the Lower Carboniferous strata are composed of grey and greyish green sandstone, siltstone, mudstone and basalt in the west, and sandstone, breccia, andesite porphyrite and rhyolite in the east.

After the breakup of Rodinia paleocontinent, the paleo-Asian ocean was initially formed in central Asia, and Junggar was a fragment of the supercontinent in the paleo-Asian ocean (Liu, 2012). In O_3, with the gradual consumption of the paleo-Asian ocean, the Early Palaeozoic continental crust was formed in the periphery of the paleo-Asian ocean, until S_1, the paleo-Asian ocean had shrunk a lot in area; meanwhile, Junggar massif became thicker gradually and a retroarc basin was formed in the trailing edge area. From S_2 to D_2, western Junggar residual ocean died out, and Balkhash-Tacheng block located in central-western Kazakhstan plate collided with Junggar block, the ophiolite discovered in the region of Daerbute faulted zone might be related to this tectonic movement. Afterward, the whole area stayed at post-collisional extension stage for a quite long period. During C_2, small scale compression occurred in the basin, and the small ocean basins in the periphery died out gradually and the basin entered into the stage of residual ocean basin, which lasted to the end of Carboniferous. Then, it entered into another ephemeral intense extensional period, when typical mafic and acidic volcanic activities occurred. In the middle and late Permian, the northwestern margin entered into foreland basin regeneration stage when overthrust faults were extensively developed. Afterwards, the basin entered upon the development stage of intracontinental depression basin, when extension and compression happened alternatively, but weak in intensity, these activities had little reformation to the structural framework in the northwestern margin of the basin (He, et al., 2006) (Figure 1-5, Table 1-1).

Table 1-1 Division of evolutionary phase of the Junggar Basin and its northwestern margin.

Evolutionary phase		Geologic era	Junggar Basin	Western and northeastern margin of the Junggar Basin
Development of superimposed basin	Intraplate compressional depression	Quaternary	The faulted block at the basement of the basin experienced bending deformation due to the loading of north Tianshan, dipping to the south	Faulted blocks, Zaire Mountain and Kalamaili Mountain etc uplifted, slipping and thrusting between the faulted blocks at the basin basement, but causing little compressional-depression deflection to the basin
		Neogene		
	Intraplate equilibrium	Paleogene	It was a uniform subsiding area, the deposited Cretaceous and Paleogine were distributed uniformly in this basin	It was a uniform uplifting area. Denudation was dominant and provided the material source for the Junggar Basin
		Cretaceous		
	Intraplate compressional depression	Jurassic	It was the compressional depression zone of strike-slip thrust basement faulted downdip block. Subsidence-sedimentation and strike-slip compressional deformation occurred	It was an uplifting zone of strike-slip thrust basement faulted updip block. Denudation was dominant, and the obduction from the north to the south was relatively weak
		Triassic		
	Intraplate rifting	Permian	It was a rift basin group. The western rift trended NNE and the northern rift trended NE-NEE. The boundary of the ancient rift basin is not consistent with the current basin trend	It was the uplifting area at the shoulder of the rift basin, and some rifting had effect on the western margin of current basin
Formation of basin basement	Plate merging	Carboniferous	Its main part was an old landmass, which might connect with Tuha old landmass at the early stage, but began to separate from it in Carboniferous, and continental marginal basins were developed in the south and the north	The northern margin was a faulted continental margin in Devonian and was transformed into an active continental margin at the late stage. Island arc-interarc basins and retroarc basins were developed at the western and eastern margins. At the late stage, the paleo-ocean basin subducted, consumed and closed
		Devonian		

Continued

Evolutionary phase	Geologic era	Junggar Basin	Western and northeastern margin of the Junggar Basin
ocean-continent separation	Silurian	Junggar-Tuha landmass was separated out due to Sinian rifting, and continental marginal basins were developed on the south and the north	It was a contract paleo-ocean basin. The ocean basin at the northeastern margin consumed during Middle Ordovician. Junggar massif and Altai massif merged and accreted to the northwestern margin of Siberia plate. The northwestern margin was a passive continental margin
	Ordovician		
	Cambrian		
	Sinian		
Formation of old landmass	Pre-Sinian	Junggar old landmass became a part of Rodinia supercontinent at the early stage of Neoproterozoic (1300–1100 Ma), and a broken piece in paleo-Asian ocean during Riphean (900–800 Ma), drifting between Siberia old land and East Gondwana land	

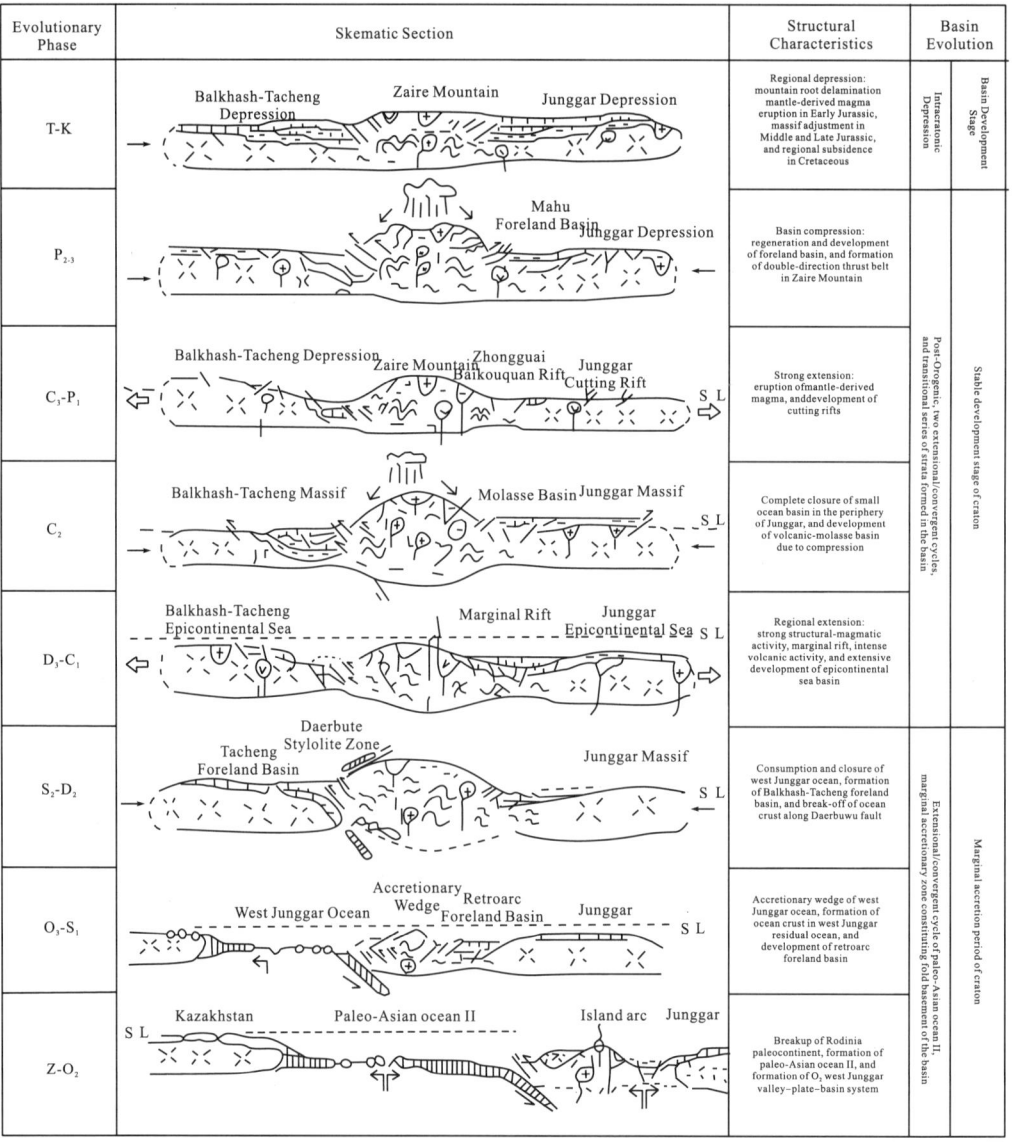

Figure 1-5 Skematic section of structural evolution in the northwestern margin of the Junggar Basin (He et al., 2006).

The formations in the northwestern margin of the Junggar Basin have the following development characteristics (Sui, 2013): ①There develop 5 regional angular unconformities, P/C, P_2/P_1, T/P, J/T and K/J from bottom to top, in which faulted or bending fold deformation are quite obvious at the planation surfaces represented by P/C, P_2/P_1 and T/P unconformities, but J/T and K/J unconformities are relatively gentle with few faults penetrating them; ②Faults are most developed in the Carboniferous-Permian, with major faults mostly cutting to the Triassic strata (or Jurassic), and there are few faults in the Cretaceous and the strata above; ③Carboniferous-Permian strata are structurally complex, and multiple "laminations" are developed in faulted zones (especially in the Hala'alate Mountain area), and Mesozoic and Cenozoic strata, structurally simple, thin quickly to the uplift area in the shape of spoon and overlie immediately Carboniferous or Permian; ④Cenozoic strata become thin quickly in the shape of wedge from the south to the north, thick Neogene strata deposit in Chepaizi area, but Cenozoic is absent in the Hala'alate Mountain and Karamay areas.

The above mentioned geological phenomena (Sui, 2013) show that the northwestern margin of the Junggar Basin experienced multiple phases of tectonic movements, in which the tectonic movements before Middle Triassic were most intensive, forming a great number of faults, folds and unconformities; before Jurassic, reworking was once remarkable, resulting in overall folding of the Triassic and the strata below, and faults were active again; Jurassic-Cretaceous strata are even with little deformation, dominated by overall uplifting and subsiding; Cenozoic strata are structurally different in the south and the north, showing obvious evolutionary differentiation. Based on the regional structural setting, basin nature and superimposed reworking characteristic in different periods, the northwestern margin of the Junggar Basin since the Permian is divided into 5 structure stages, post-orogenic extension (P_1), strong compression overthrust (P_{2+3}), inherited overthrust superimposition (T), overall oscillation and fluctuation (J-K) and intracontinental foreland (KZ). The formation of Hala'alate Mountain is the result of multiple structure superimposition. At the end of Carboniferous, small scale overthrusting happened, but the major thrusting began in the middle-late Permian, became strong during late Permian-early Triassic, and then stayed stable until late Triassic-early Jurassic when the tectonic movement reached climax, afterwards, the tectonic movement weakened, and the area moved into Yanshan-Himalayan adjustment period.

Chapter 2

Reservoir Petrological Characteristics and Reservoir Space Types

2.1 Reservoir petrology

The volcanic reservoir is a type of special oil and gas reservoir with some porosity and flowing capacity, formed from volcanic magma under the action of condensation, diagenesis and various secondary diagenesis. There are many types of volcanic rocks and corresponding many classification program, but in current academic research and practical application, the rock texture-genesis classification program recommended by IUGS (International Union of Geological Sciences) in 1989 is commonly used. Based on this program, volcanic reservoir rocks are classified into volcanic lava and volcaniclastic rock. It can be seen from the volcanic oil and gas reservoirs discovered around the world (Othman et al., 2010; Rohrman, 2003; Rushdy et al., 2002; Schutter, 2003) that volcanic reservoirs have multiple types of rocks. As volcanic reservoirs have no specific rock types (Zhao et al., 2004), not only volcanic lava (e.g., basalt, andesite and rhyolite), but also volcanic breccia (e.g., volcanic agglomerate, volcanic breccia and tuff) have the potential to become oil and gas reservoirs.

The study area had strong volcanism in Carboniferous, so the Carboniferous system there has both volcanic lava (andesite and basalt) and volcaniclastic rock (volcanic breccia and tuff), in which the volcaniclastic rock has better primary reservoir conditions and suffered stronger reformation by later tectonic movements and dissolution as well as secondary diagenesis, so it is the most important type of oil and gas reservoir rock in this area.

2.1.1 Volcanic breccia

Volcanic breccia is made up of the clastics formed during volcanism and the normal sediments or lava materials (cements) by means of accumulation and consolidation. It is mostly grey or brown grey (oil bearing), with a breccia (clastics) content of no less than one third (generally higher than 50%); the breccia varies from 2 mm to 64 mm in size; about 50 mm at maximum in the study area, breccia is mostly andesitic breccia, basaltic breccia and rhyolitic breccia. Between breccias, it is generally welded breccia and lava, and when the lava is dissolved later, forming calcite, calcite cemented volcanic breccia will come about.

In the study area, volcanic breccia formation is found in nearly all wells, with a thickness of about 300 m in general, but up to 1700 m at maximumin Well HQ3 and HQ101. In areal, it thins gradually from the northwest to the southeast along the connection line between Well HQ3 and Well HQ102. Volcanic breccia is mostly alternate with tuff, forming an eruption cycle from explosive facies to volcanic sedimentary facies (Figure 2-1).

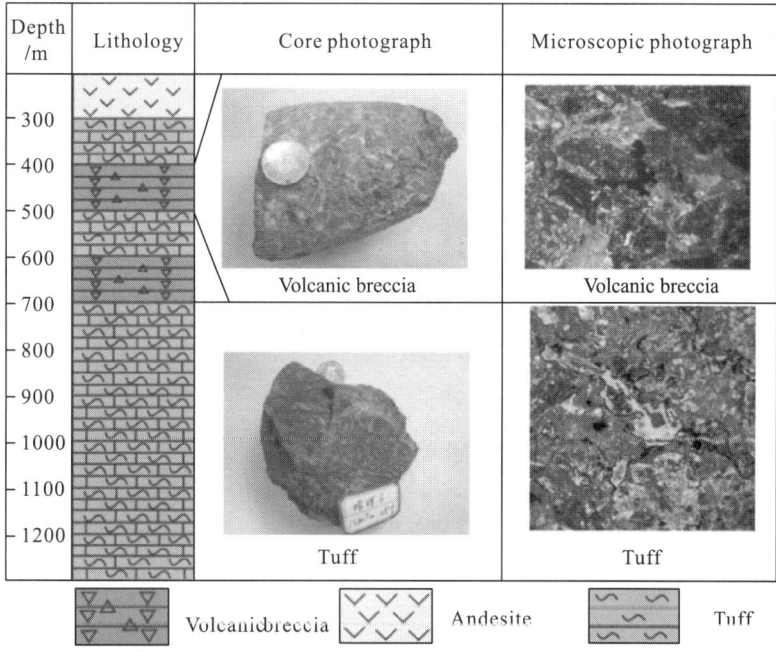

Figure 2-1 Rhythmic cycle of volcanic breccia and tuff (Carboniferous, Well HQ6).

Based on hand speciman observation, some volcanic breccia in each eruption cycle has graded structure, with normal grading and an obvious interface between large-sized volcanic agglomerates and small-sized volcanic breccias (Figure 2-2a). The volcanic breccias are mostly greyish white, and subangular, with a maximum flat side diameter of about 10 mm; flat side of most breccias are parallel to or slightly intersected with the breccia sequence bedding plane; the space between breccias is filled with dark grey volcanic ash. Welded breccias are, on the whole, in the form of block structure, in which rigid linen andesibasaltic debris is cemented with white andesitic plastic lava, the rigid debris is round with a diameter of about 25 mm, and is dissolved into embayment by andesitic lava and transited gradually into andesitic lava. There are pores filled with black bitumen inside the welded breccias (Figure 2-2b).

Volcanic breccias, obvious interface between volcanic agglomerates and breccias, Well HQ6, 812.10 m.

Welded breccias, Well HQ6, 648.50 m, Bit: bitumen. Bre: breccias.

Figure 2-2 Volcanic breccia (Carboniferous, Well HQ6).

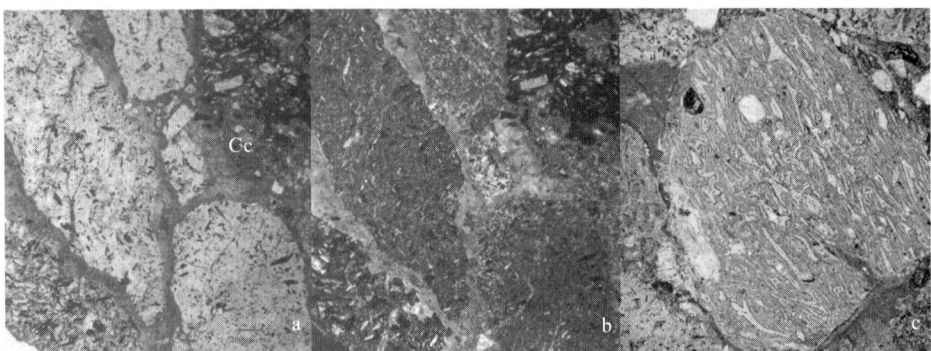

Volcanic brecciated texture, calcite cement, Well HQ6, 647.60 m, plane-polarized light, Cc: calcite.

Volcanic brecciated texture, calcite cement, Well HQ6, 647.60 m, cross-polarized light, Cc: calcite.

olcanic breccia, rhyolitic breccias, Well HQ102, 1340.77 m, plane-polarized light.

Volcanic breccia, rhyolitic breccias, Well HQ102, 1340.77 m, cross-polarized light.

Volcanic breccia, vesicle bearing hyalopilitic texture, Volcanic brecciated texture, Well HQ6, 814.10 m, plane-polarized light.

Volcanic breccia, vesicle bearing hyalopilitic texture, Volcanic brecciated texture, Well Hq6, 814.10 m, cross-polarized light.

Figure 2-3 Microscopic characteristics of Carboniferous volcanic breccias in the Hala' alate Mountain area.

Based on microscopic observation, the rocks mostly have typical volcanic breccia texture (Figure 2-3a, b, c and d) and vesicle bearing hyalopilitic texture (Figure 2-3e and f). The volcanic breccias have a long axis of 3-5 mm and length-width ratio of about 3∶1 in general. The breccias join into frames, in which fine volcanic ash fills. In later evolution, the volcanic ash is dissolved by alkaline fluid and replaced by automorphic and hypautomorphic calcite, the breccias are dissolved by fluid in angles, turning into the shape of subangular-subround, and the fractures and edges of breccias are filled with calcite. The breccias are mostly basaltic-andesitic breccias and rhyolitic breccias (Figure 2-3b), hypautomorphic feldspathic crystals replaced with calcite are frequently seen in basaltic-andesitic breccias, and stretched vesicles are orientatedly arranged in rhyolitic breccias, filled with late quartz, and in the shape of chicken bones (Figure 2-3c and d). In vesicle bearing hyalopilitic texture, vesicles, mostly round, are disorderly scattered in matrix with a content about 20% and diameter of 0.1-0.3 mm, and filled with quartz and automorphic calcite (Figure 2-3f); in the matrix, automorphic-hypautomorphic feldspar particles are interlocked with fine volcanic ash and vitroclastic pariticles, into hyalopilitic texture. As volcanic breccia is mostly formed by virtue of volcanic explosion, near volcano crater, so dissolved pores, vesicles and fractures are developed in volcanic breccias.

2.1.2 Andesite

Andesite is a kind of neutral-basic extrusive rock in the category of calc-alkaline series, and is silicic acid saturated and weakly saturated. It is greyish-white, with porphyritic texture and blocky structure.

Figure 2-4 Multiphase cycle of andesite and tuff(Carboniferous, Well HS1).

In Wells HS1, HS2, HQ3 and HQ6, andesite intervals with different thickness have been found.

The andesite layer is thickest in Well HS1 and HS2, up to 800 m; while multiphase cycles of andesite (600 m) and tuff (1000 m) are presented in Well HS1 (Figure 2-4), indicating this well is located at the distal end of the volcano. In Well HS2, andesite (800 m), tuff (800 m) and basalt (900 m) form multiple cycles, indicating the well is relatively close to the volcano crater. In Well HQ6, andesite (100 m), tuff (800 m) and volcanic breccia (300 m) form multiphase cycles, indicating this well is close to the volcano crater.

Core observation shows that the andesite in the above-mentioned well intervals, in blocky structure, has fine columnar feldspar and lots of dark minerals visible by naked eye, rich fractures and pores, some of them filled with calcite. Calcite veins are distributed irregularly. A small amount of black bitumen in sparse star is seen at the cross section of cores.

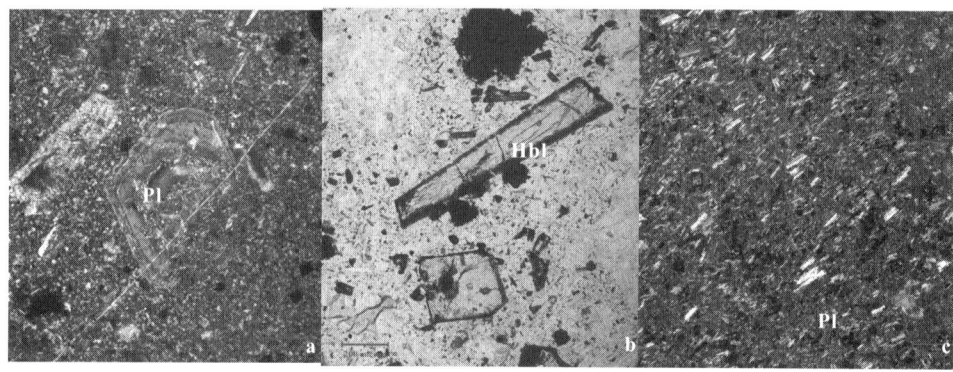

Ring texture, dissolution structure and porphyritic texture of plagioclase of basaltic andesite, Pl: plagioclase (intermediate), Well HQ7, 160.60 m, cross-polarized light.

Opacitization edge of hornblende of andesite, Well HQ7, 162.65 m, plane-polarized light.

Hyalopilitic texture of andesite, Well HQ7, 213.15 m, cross-polarized light, Hbl: hornblende, Pl: plagioclase micrite.

Figure 2-5 Microscopic characteristics of Carboniferous andesites in the Hala' alate Mountain area.

Based on thin section analysis, the andesite is mostly of porphyritic texture (Figure 2-5a), in which phenocryst is mainly composed of plagioclase in the shape of thick board, long columnar and tuberose (15%–20%), and dark minerals. Mainly andesine and labradorite, plagioclase, astatic, usually has ring texture and dissolution structure caused by intense alteration (Figure 2-5a). The dark minerals are primarily hornblende and occasionally pyroxene or biotite, in which most hornblende grains have opacitization edge structure (Figure 2-5b), pyroxene has simple twin crystals, and biotite grains are mostly dissolved, forming secondary pores. Andesite matrix has hyalopilitic texture (Figure 2-5c), and plagioclase micrites are in interlocking or parallel arrangement with a small amount of hyaline; there is minor quartz, with little yellow edge dissolved by matrix; there are vesicles filled with quartz and calcite in the middle, showing multiphase filling feature.

Chapter 2 Reservoir Petrological Characteristics and Reservoir Space Types

Based on core observation and thin section analysis, the andesite, affected by structural fractures, commonly has fracture-type pores, which are good reservoir space, and often contain oily fillings.

2.1.3 Basalt

Basalt is a kind of basic extrusive rock in the category of calc-alkaline series, and is silicic acid unsaturated and weakly saturated. It is greyish black with blocky structure.

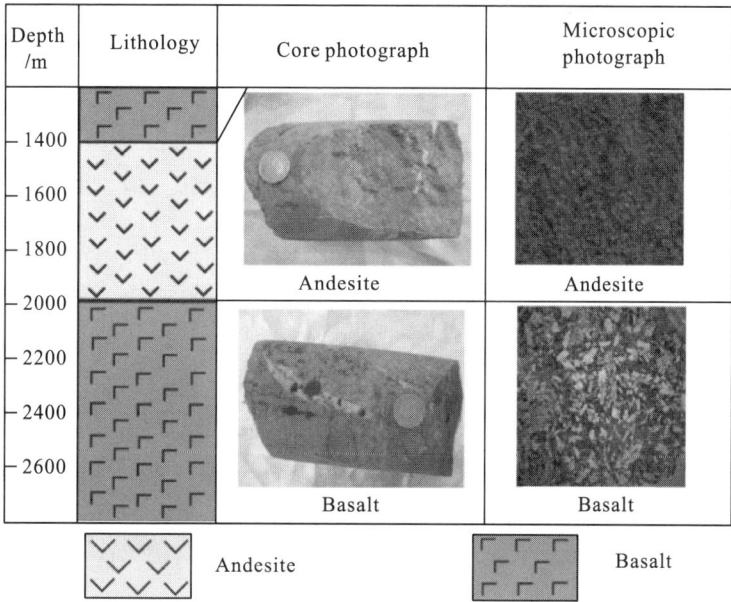

Figure 2-6 Multiphase eruption cycle of andesite and basalt (Carboniferous, Well HS2).

Basalt intervals are discovered in Well HS2, HSE2 and HQ4 in the study area. The basalt in each cycle is not uniform in thickness, indicating that the volcano eruption of different periods are different in intensity and nature. The basalt in a cycle is up to 700 m thick. In Well HS2, basalt (900 m), andesite (800 m) and tuff (800 m) constitute multiphase cycles, showing that basalt is relatively far away from the crater (Figure 2-6), and multiphase eruption cycles from basic to intermediate rocks.

Based on core observation, the basalt in above mentioned well intervals, greyish black-brownish black, has vesicle, amygdaloidal and blocky structures (Figure 2-7a). The basalt core samples have large amounts of dark minerals, abundant fractures and pores, in which the pores are mostly primary pores filled with calcite, quartz or ferruginous materials; fractures are mostly filled with calcite vein or quartz vein. The veins, distributed disorderly, have a small amount of black bitumen.

Based on microscopic section observation, the basalt, mostly trachybasalt texture, has granular pyroxenes and fine magnetite particles filling the gaps formed by irregularly distributed band plagioclase micrites (Figure 2-7b and c), indicating that the basalt was formed at lower cooling speed. Dark minerals are primarily pyroxene and occasionally hornblende with opacitization edge texture. Affected by structural fractures, the basalt has fractured pores, which can be fairly good reservoir space, and the fractures are frequently filled with oily materials.

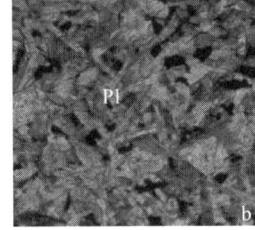

Dark grey basalt, amygdaloidal structure, Well HS2, 2024.43 m (core photograph).   Basalt, trachybasalt texture, Well HQ4, 686.50 m, plane-polarized light.   Basalt, trachybasalt texture, Well HQ4, 686.50 m, cross-polarized light.

Figure 2-7 Characteristics of Carboniferous basalt in the Hala'alate Mountain area.

2.1.4 Tuff

Tuff is formed by volcanic ash grain size of less than 2 mm accumulating at the surface after floating in the air during volcanic eruption. Due to its small grain size, volcanic ash can fly tens even hundreds of kilometers after blasted off crater, so it is commonly accumulated far away from the crater. During ascending, volcanic ash grains weld with each other, forming welded tuff, which is generally tighter with pseudofluidal structure and occasionally beddings.

In Well HS1, HS2, HQ3, HQ6, HQ7, and HQ102, tuff core intervals of different thickness have been discovered (Figure 2-8). The tuff interval is the thickest in Well HS1, up to 1200 m. Multiphase eruption cycles of tuff and andesite are presented in Well HS1 (Figure 2-4), indicating the well is far from the volcanic crater. In Well HS2, tuffs, andesite and basalt constitute multiple cycles (900 m), indicating this well is relatively close to the volcanic crater; and in Well HQ6, tuff, andesite and volcanic breccia form multiphase cycles, proving this well is close to the volcanic crater.

Core samples taken from the above intervals show the tuff is generally greyish green, tight and brittle, with rare lithic and crystal invisible to the naked eye. There is a small amount of lithic, vitrieous fragment and crystal in the tuff, proving that the tuff is the product of volcanic explosion, and volcanic sedimentary facies.

Secondary dissolution often happens in the tuff, forming dissolved pores in matrix (Figure 2-9). Calcitization, silication and chloritization are often seen in the dissolved pores, indicating multiphase dissolution and refilling. Fractures were filled with calcite later, which was dissolved again, giving rise to pores with better reservoir properties.

Chapter 2 Reservoir Petrological Characteristics and Reservoir Space Types

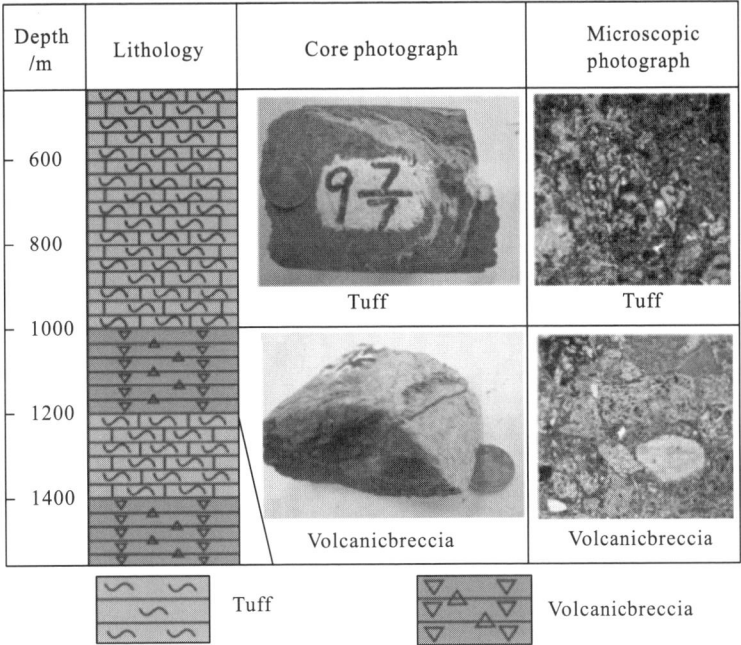

Figure 2-8 Multiphase eruption cycle of volcanic breccia and tuff (Carboniferous, Well HQ102).

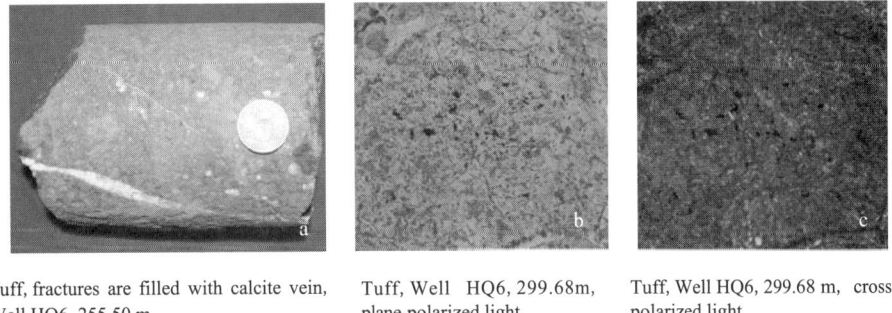

Tuff, fractures are filled with calcite vein, Well HQ6, 255.50 m.

Tuff, Well HQ6, 299.68m, plane-polarized light.

Tuff, Well HQ6, 299.68 m, cross-polarized light.

Figure 2-9 Characteristics of Carboniferous tuff in the Hala'alate Mountain area.

2.1.5 Clastic rock

Clastic rock in the study area was developed from the Permian to the Cretaceous, but the Triassic strata are commonly absent in drilled wells in the study area. Seismic sections show that the sedimentary formations in the study area are, on the whole, dipping to the south, clastics rock formations are in unconformable contact with the underlying Carboniferous strata, and 100 m to 300 m thick in the drilled wells. Fairly complex in lithology, the rock types in the study area include fine conglomerate (Figure 2-10a and d), siltstone (Figure 2-10b), oil flecked mudstone (Figure 2-10c) and oil flecked sandstone (Figure 2-10e). There are oil and gas shows in the clastic rock mainly in the form of oil and bitumen, which are mostly distributed in the fissures and beddings in clastic rock and in the pores between conglomerate grains.

The clastic rock in the study area is mainly dark grey and linen, with clastic texture and blocky structure. Fairly good in sorting and roundness, with clastic texture and blocky structure, sandstone and siltstone are mainly composed of clastic particles and matrix, and the clastic particles comprise feldspar, quartz and lithic particles. Fine conglomerate, better rounded, have gravels 20–30 mm in diameter, at maximum 30 mm. Clastic rock has rhythmic beddings in general (Figure 2-10e), and pores and fractures mostly filled with calcite or quartz (Figure 2-10f). In the clastic rock, oil and bitumen are mostly distributed in the pores between clastic particles or conglomerate particles and in the beddings of the rock, with good oil and gas shows.

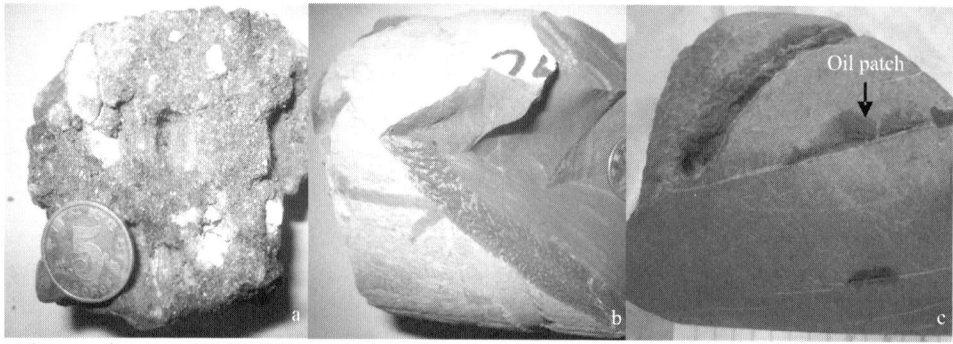

Fine conglomerate, Well HQ4, 551.59 m.   Siltstone, well HQ6, 1360.70 m.   Silty mudstone, Well HS1, 2098.90 m.

Fine conglomerate, clastic texture, Well HQ4, 551.59 m, cross-polarized light.   Siltstone, rhythmic bedding, Well HS1, 2094.70 m, cross-polarized light.   Siltstone, fractures filled with quartz, Well HS1, 2154.50 m, cross-polarized light.

Figure 2-10 Characteristics of Permian clastic rock in the Hala'alate Mountain area.

The mudstone is harder, but finer fractures are observed in the mudstone cores, and oil and gas shows occur in the fractures, the section of fractures with oil are brown, while the fractures without oil are grey.

Based on thin section observation, siltstones and sandstones in this study area are mainly of psammitic texture and blocky structure, and rhythmic bedding is developed in some sandstone samples. Mineral components in the sandstone mainly include feldspar, quartz and lithic, and a small amount of mica. The quartz, 50%–60% in content, is smaller in grain size under cross-polarized light, and exist in crystal grain aggregates, and occasionally in the form of veins. Larger quartz particles have better roundness and sorting. The feldspar with a lower content of about 10%, has polysynthetic twinfeature under cross-polarized light. In addition, lithic particles are also identified in the thin sections with a content of about 30%, and are mainly sedimentary lithic and volcanic lithic (e.g., basalt lithic and volcanic tuff lithic). Mica, accounting for 5%, mainly consists of muscovite and sericite, and under cross-polarized light, mica appears as scale aggregates with high interference color and bright color, so it can be distinguished easily. They are all matrix except clastic particles. Under plane-polarized light, the gravels are of clastic texture, various in size, generally in the range of 20–30 mm (at maximum about 30 mm). The gravels are poorly sorted, with roundness degree in between subrounded and rounded, and mainly comprise sandstone, siltstone and volcaniclastic rock. The pores between gravel particles are filled with matrix and quartz particles in angular shape. It can be seen clearly by microscopes that the pores between clastic particles are filled with bitumen and oil, and oil and gas shows also can be seen in the beddings of sandstone.

Mudstone in this study area also has fairly good oil and gas shows. Under cross-polarized light, the mudstone, blocky structure, has horizontal beddings and fine fractures, and there is bitumen in the fractures.

2.2 Reservoir space types

Diverse in type, complex in genesis and strong in heterogeneity (Zou et al., 2008), the development and distribution of volcanic reservoir space are affected by multiple geologic factors. Volcanic rock generally has primary pores and fractures during initial consolidation and diagenesis, afterwards, a large amount of secondary pores and fractures would form after multiphase tectonic activities, weathering leaching and dissolution of formation fluids in the volcanic rock (Wang et al., 2008; Li et al., 2008). With primary reservoir space of limited scale and poor connectivity, volcanic rock can hardly become good oil and gas reservoirs without secondary reworking at the later stage. In the area of the Hala'alate Mountain, the northwestern margin of the Junggar Basin, Xinjiang, the Carboniferous volcanic reservoirs have multiple types of reservoir spaces, including primary vesicle, interbreccia pore and intracrystalline blast fracture, and secondary intracrystalline dissolved pore, matrix dissolved pore and structural fracture etc, in which the structural fractures of secondary genesis and the later dissolved pores, cavities and fractures play vital roles in making the volcanic rock there good reservoir.

Through observation of cores from Well HS1, HS2, HQ3, HQ101, HQ102, HQ4, HQ6 and

HQ7, and analysis of their thin sections, according to the types and features of macro- and micro- reservoir space in the volcanic rock in the study are, and the genesis of various reservoir space types, the classification scheme for the Carboniferous volcanic reservoir space in this study area has been worked out: firstly, the reservoir space is classified into primary reservoir space and secondary reservoir space according to genesis; then based on reservoir space morphology, the primary reservoir space is divided into vesicle, interbreccia pore, condensed shrinkage fracture and intracrystalline blast fracture, and the secondary reservoir space is divided into intracrystalline dissolved pore, matrix dissolved pore, dissolved pore in vesicle filler and structural fracture (Table 2-1).

Table 2-1 Reservoir space types in the Hala'alate Mountain area and their characteristics.

	Pore type	Pore genesis	Pore characteristic	Representative lithology
Primary reservoir space	Vesicle	Volatile components which don't escape in time are preserved in the rock, forming vesicles	Various in morphology, including round-oval, linear or irregular shapes	Andesite, basalt
	Interbreccia pore	Residual pores between volcaniclastic particles after diagenesis	Irregular in morphology, commonly linear	Volcanic breccia
	Condensed shrinkage fracture	Constant-volume cooling	Irregular or banded morphology	Basalt, andesite
	Intracrystalline blast fracture	Crystal condensation, shrinkage and blast	Irregular morphology, cutting crystals or growing along cleavage	Volcanic breccia
Secondary reservoir space	Intracrystalline dissolved pore	Partial dissolution inside crystals	Irregular morphology, inside crystals	Volcanic breccia, andesite
	Matrix dissolved pore	Devitrification of scorilite or dissolution of micritic feldspar	Fine mesh shape	Volcanic breccia
	Dissolved pore in vesicle filler	Recrystallization or partial dissolution of vesicle filler	Irregular morphology	Andesite
	Structural fracture	Fractures induced by tectonic stress	Straight and good extension	Tuff volcanic breccia

2.2.1 Primary reservoir space

Primary reservoir space refers to various types of reservoir space formed in volcanic rock during the condensing diagenesis of magma before secondary tectonic and fluid diagenesis (Xiong et al., 2012). Primary reservoir space in volcanic rock is affected by multiple factors, including chemical composition of magma, volatile components and contents, diagenesis patterns, volcanism intensity and topography during volcanism. Primary pore in volcanic rock mainly refers to the nonlinear reservoir space formed and preserved during the consolidation diagenesis of volcanic rock, which include vesicle and interbreccia pore in the study area. Primary fracture in volcanic rock refers to linear fractured reservoir space formed due to condensation shrinkage during the consolidation diagenesis of volcanic rock, which includes two types, condensed shrinkage fracture and intracrystalline blast fracture in the study area.

1. Vesicle

The volume of volatile components in magma increases with the decrease of pressure, when magma flows at the surface after erupting out of ground; and owing to buoyancy, volatile components get out of rock successively, but part of volatile components fail to escape are preserved in magma, forming vesicles. There are sustantial primary vesicles in extensive distribution in this area. The vesicles are diverse in shape, including round, oval and irregular ones. A small amount of vesicles are elongated and oriented at different degrees along the flowing direction of magma. The vesicles are various in size and density, and poor in connectivity. This type of vesicle is mainly distributed in the middle and upper sections of volcanic lava flow, and is directly related to the content of volatile components in magma. Generally, there are more vesicles in neutral magma than in basic magma (Figure 2-11a and d).

Volcanic breccia is mostly explosive facies close to the volcano where magma eruption is intensive, so there are a great number of vesicles of different sizes in volcanic breccia, and most of them are filled with carbonate and kiesel of later stage. The vesicles in andesite are formed when the gas escapes out of lava with the decrease of pressure in the process of condensation because the temperature and pressure change sharply when magma erupts out of the strata, and they are different in morphology and size. Magma vesicles are mostly distributed at the upper section of andesite magma, and decrease in number gradually downwards, and they are mostly independent of each other, but connected with secondary fractures and dissolved pores and fractures. Moreover, they are often filled with hypautomorphic-allotriomorphic opaque metallic mineral and recrystallized fine felsic minerals, forming amygdaloids.

2. Interbreccia pore

The primary interbreccia pores in volcanic rock in this area mainly include two types: one, the interbreccia pore formed by cryptoexplosive brecciation or in volcanic lava clastic breccias (Figure 2-11b and e), but this type of pores are mostly filled or partially filled; the other, the intervolcaniclastic pore in volcaniclastics, which is generally larger and well connected, often associated with fractures, interbreccia pores are controlled by the combination patterns between volcaniclastics and volcanic breccias in size and shape.

3. Intracrystalline blast fracture

The crystals formed earlier in magma could break to form fractures (intracrystalline blast fractures), because the confining pressure drops quickly and the pressure of magma itself decreases sharply in the process of volcanic eruption. Intracrystalline blast fractures are generally irregular in morphology, and often cut through crystals, allowing later fluid to flow along and dissolve the crystals. Intracrystalline blast fractures commonly occur in the rock with porphyritic texture and fine volcanic breccia with crystals (Figure 2-11c and f).

Primary reservoir space is universal but uneven in distribution in the study area, and different types of rocks have different primary reservoir space combinations. In volcanic breccia, there are primary interbreccia pores and condensed shrinkage fractures; in basalt and andesite, there are vesicles and intracrystalline blast fractures, and in comparison, volcanic breccia has higher primary porosity.

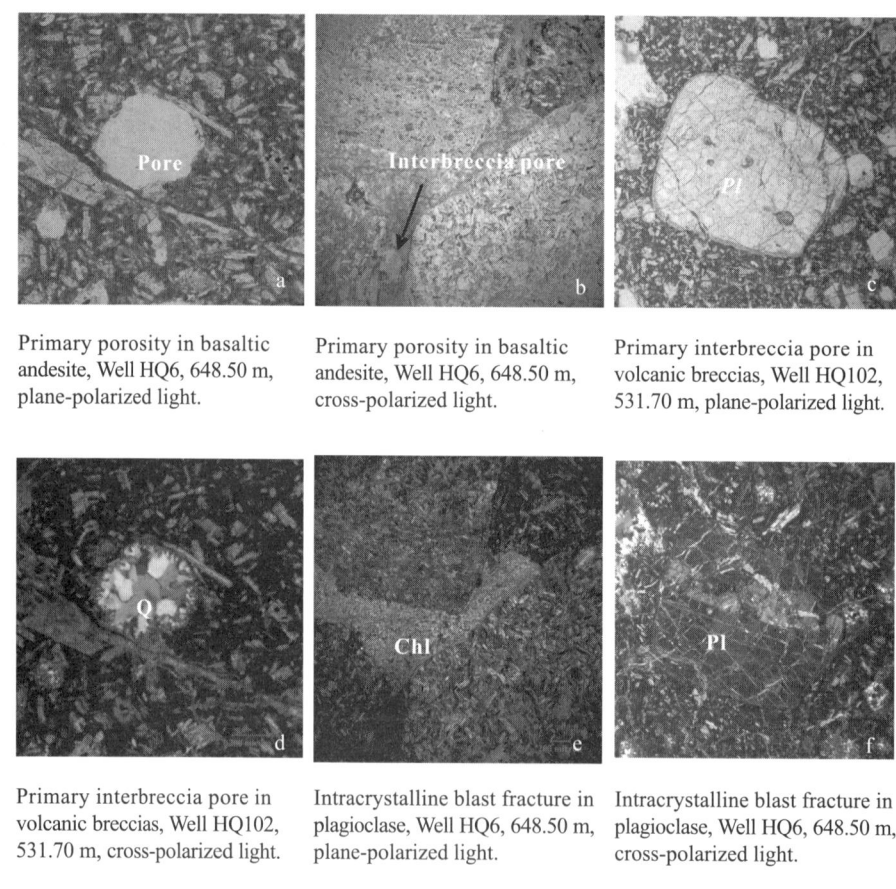

Primary porosity in basaltic andesite, Well HQ6, 648.50 m, plane-polarized light.

Primary porosity in basaltic andesite, Well HQ6, 648.50 m, cross-polarized light.

Primary interbreccia pore in volcanic breccias, Well HQ102, 531.70 m, plane-polarized light.

Primary interbreccia pore in volcanic breccias, Well HQ102, 531.70 m, cross-polarized light.

Intracrystalline blast fracture in plagioclase, Well HQ6, 648.50 m, plane-polarized light.

Intracrystalline blast fracture in plagioclase, Well HQ6, 648.50 m, cross-polarized light.

Figure 2-11 Characteristics of primary pores in Carboniferous volcanic rock in the study area.

4. Condensed shrinkage fracture

After erupting out of the strata, lava flows and condenses at different rate, in the process condensed shrinkage fractures are formed. Tensile, with irregular occurrence and generally smaller in scale, condensed shrinkage fractures often appear in the volcanic magma with flow structure and in the matrix of volcaniclastic with lava texture. The microfractures in andesite are mostly generated by condensing shrinkage of magma or recrystallization of scorilite, distributed along the edges of crystals and some connected with dissolved pores.

2.2.2 Secondary reservoir space

Occurrence and results of secondary diagenesis are largely dependent on the characteristics of the primary reservoir. Secondary reservoirs are often superimposed on primary reservoirs, leading to complicated pore types in volcanic reservoirs (Liu et al., 2010). Secondary diagenesis has dual effects on volcanic reservoir properties, on one hand, it causes complete or partial filling of primary pores, lowering the reservoir capacity of volcanic rocks to some extent; on the other hand, it causes the breakup of volcanic rock at different degrees, giving birth to a large amount of secondary pores and enhancing storage capacity of volcanic rock. The development of secondary pores is favorable for the accumulation of oil and gas (Liu et al., 2012), in the study area, volcanic reservoirs with developed secondary pores often have good oil and gas shows.

The volcanic rock in the study area was mainly generated in the Carboniferous, and has experienced multiphase tectonic movements, including Hercynian, Indosinian, Yanshan and Himalayan since then (Hu, 2012). The primary reservoir space has been reworked under the joint effect of multiple diagenesis, including compaction, solution, recrystallization and secondary alteration, as a result, intracrystalline dissolved pores, intramatrix dissolved pores and fractures, dissolved pores in vesicle filler and structural fractures have been formed.

1. Intracrystalline dissolved pore

The individual minerals often seen in volcanic rock in the study area include feldspar, quartz, pyroxene, hornblende and biotite. Among these mineral crystals, only quartz crystals are chemically stable, while the rest often exchange composition with formation fluid, and the unstable mineral components are dissolved due to dissolution and hydrolysis, so mineral crystals are structurally changed and dissolved pores are formed where crystals are dissolved. In the study area, intracrystalline dissolved pores in andesite are mainly developed in feldspathic individual crystals (Figure 2-12), mostly neutral plagioclases, due to various types of secondary change, including sericitization, clayization, epidotization and calcitization, and the intracrystalline dissolved pores are mainly filled with calcite, oil and bitumen.

2. Intramatrix dissolved pore and fracture

When the geologic conditions are satisfied, alteration and dissolution, such as chloritization, zeolitization and scorilite devitrification, take place in the matrix of volcanic rock, and the products of these secondary diageneses are dissolved by formation fluid, giving rise to secondary dissolved pores (i.e., intramatrix dissolved pores). Generally smaller, but large in quantity, and connected to some extent, they are the principal reservoir space in tuff. In the study area, dissolved pores are developed in volcanic breccia, in the matrix, a great number of large dissolved pores are commonly seen. This kind of pore usually has the characteristic of multiphase filling, with stage one siliceous fluid giving rise to comb-edge quartz along the inner wall of dissolved pores, and stage two siliceous fluid giving rise to individual-crystal automorphic quartz filling the remaining space of the pores (Figure 2-13). It is shown in Figure 2-14 that the edges of feldspar is smooth, and feldspar in the matrix is dissolved, forming ring edges. Tuff matrix is melted into bay or round shape by magma upwelling from deep formation which carries high-temperature fluid when reaching the shallow formation or erupting out of the surface, and as a result, intramatrix dissolved pores are formed (Figure 2-14).

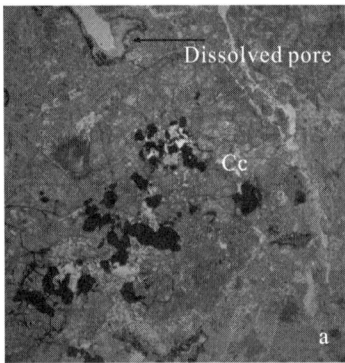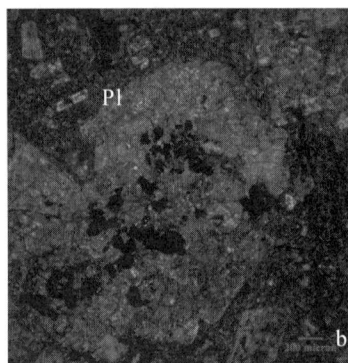

Figure 2-12 Dissolved pores in feldspathic crystals filled with later calcite and bitumen.214.00 m in Well HQ7, Cc: calcite, Pl: plagioclase, a: plane-polarized light, b: cross-polarized light.

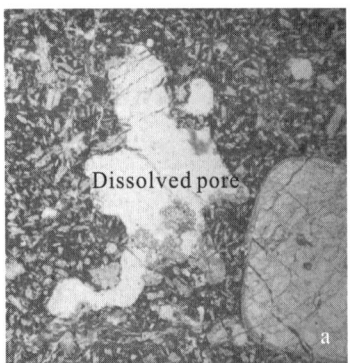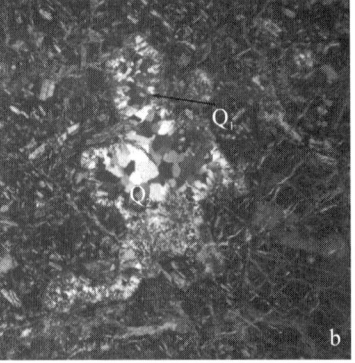

Figure 2-13 Matrix dissolved pores in volcanic breccia. Q_1: stage one siliceous filling in dissolved pores (comb edge), Q_2: stage two siliceous filling in dissolved pores (automorphic quartz), 648.50 m in Well HQ6, a: plane-polarized light, b: cross-polarized light.

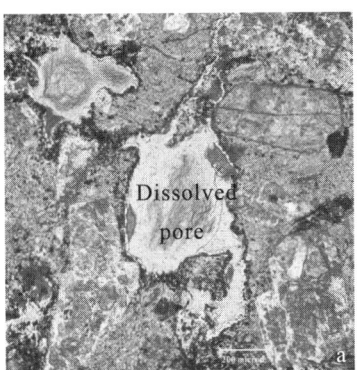 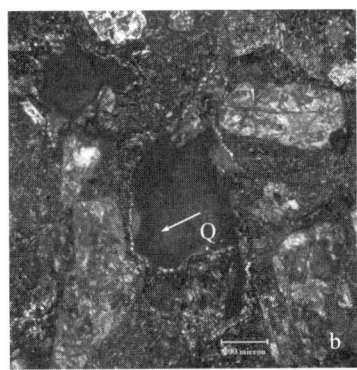

Figure 2-14　Dissolved pores in the matrix of tuff (filled with stage-one siliceous at the edge). 214.00 m in Well HQ7, Q: quartz, a: plane-polarized light, b: cross-polarized light.

In the study area, volcanic matrix dissolved fractures are commonly seen in the Carboniferous volcanic breccias. This type of fracture is derived from dissolution of existing structural microfractures, when the volcanic matrix is dissolved by the later formation fluid. The matrix dissolved fractures in the study area are long, wide and dense, and are often filled with siliceous, carbonate and crude oil etc (Figure 2-15). Capable of connecting primary pores, matrix dissolved fractures are conductive to the improvement of oil and gas flowing and the storage capacity of volcanic breccia.

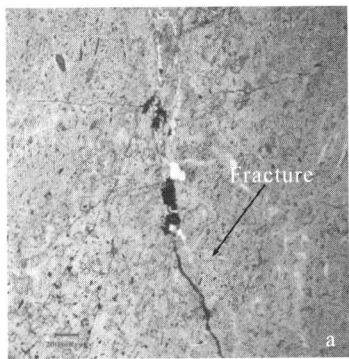 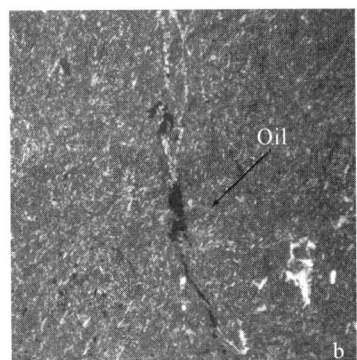

Figure 2-15　Dissolved fractures in the matrix of volcanic breccia (oil bearing). 302.81 m in Well HQ6, a: plane-polarized light, b: cross-polarized light.

3. Dissolved pore in vesicle fillings

Based on core observation, fillings in vesicles mainly includes zeolite, chlorite, bitumen, calcite and silica. Secondary pores (i.e., dissolved pores in vesicle filler) are formed when the minerals inside vesicles are completely or partially dissolved (Figure 2-16). This type of pore is mainly found in volcanic lava. It is shown in Figure 2-16 that primary vesicles in basaltic andesite were filled with secondary calcite, which was dissolved later, forming new dissolved pores.

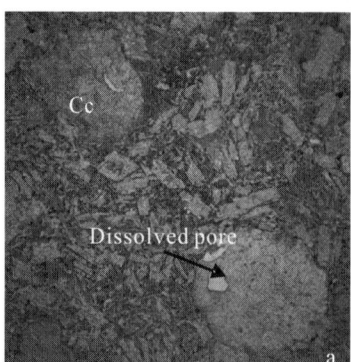 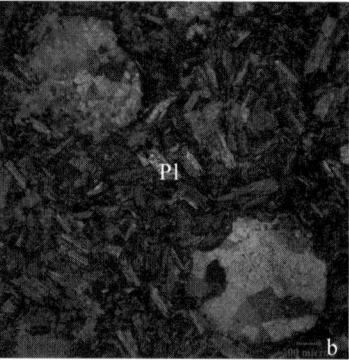

Figure 2-16 Dissolved pores in vesicle filler of basaltic andesite. Vesicles were filled with calcite after formed, and then new pores are formed due to the dissolution of calcite at later stage; Cc: calcite, Pl: plagioclase, 1339.04 m in Well HQ102, a: plane-polarized light, b: cross-polarized light.

4. Structural fracture

Volcanic rock is tight and brittle with low primary porosity and permeability, it is likely to form fractures under the action of tectonic stress (Ruan et al., 2012). The fractures in volcanic rock not only act as important fluid flowing pathways, but also as good reservoir space. The development of fractures plays a controlling role in the formation of dissolved pores and cavities later. Therefore, fractures are one important part for volcanic reservoir study, and the principal target of oil and gas exploration and development (Gao and Xie, 2007; Dai et al., 2003). Generally, structural fractures often appear in batches, in the form of a group of high-angle and low-angle fractures of a certain length, or tiny fracture systems in local areas. Structural fractures, as important migration pathways of various formation fluids, improve the permeability of volcanic rock significantly (Qu, 2015). That is to say that the volcanic rock with developed structural fractures would suffer stronger later diagenesis.

Multiphase tectonic movements happened in the study area (Hu and Xia, 2012), and a great number of structural fractures have been formed in the Carboniferous-Permian volcanic rock due to strong nappe activities. Observation of hand specimen and thin section shows the structural fractures in the study area are often filled and reworked by the later diagenesis in the form of complete or partial filling with kiesel and calcic. Secondary fractures are the products of matrix dissolution by fluid or tectonism, are wide, long, and dense, and often filled with siliceous, carbonate and crude oil (Figure 2-17). This type of fracture, connecting primary pores, is vital for volcanic breccia to become good oil storage space. Developed fractures in tuff are formed by tectonism, and mostly filled with calcite (Figure 2-17), and the calcite is dissolved at later stage, forming good reservoir space.

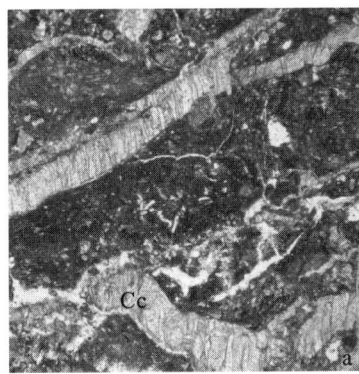 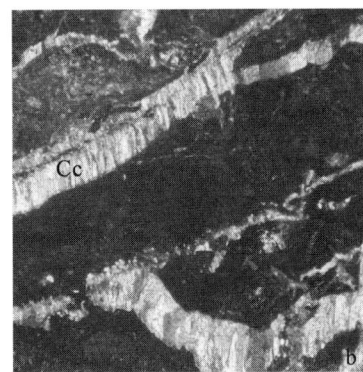

Figure 2-17 Fractures filled with calcite veins in tuff. 257.80 m in Well HQ6, Cc: calcite, a: plane-polarized light, b: cross-polarized light.

The formation and development of secondary pores in volcanic reservoirs are usually influenced by the nature of primary reservoirs. Secondary reservoirs are often superimposed on primary reservoirs, leading to complex pore types in volcanic reservoirs. Secondary diagenesis has dual effects on reservoir properties of volcanic rock, on one hand, primary pores are completely or partially filled, lowering the reservoir quality of volcanic rock to some extent; on the other hand, secondary diagensis causes breakup of volcanic rock to various degrees, and thus the formation of a large amount of secondary pores, improving the reservoir quality of volcanic rock. The storage and flowing capacities of volcanic reservoirs in the study area are reworked by secondary diagenesis, and fractures play a constructive role in the reservoir quality of the volcanic rock.

2.3　Factors influencing reservoir space development

Based on previous studies, the development of volcanic reservoirs in the Junggar Basin is influenced by multiple factors (Zhang et al., 2015; Liu et al., 2012; Li et al., 2008; Zhao and Shi, 2012), mainly including volcanic eruption environment, lithology and lithofacies, and the later weathering leaching and fracturing reworking. For the Carboniferous-Permian volcanic reservoirs in the study area, regional tectonism is the basic condition for the development of volcanic rock, lithology and lithofacies control the development of primary reservoir space in the volcanic rock, and diagenesis and later tectonism are the key factors to further improve volcanic reservoir quality later.

2.3.1　Effect of lithology and lithofacies

Volcanic facies refers to the formation conditions of volcanic rock and the corresponding lithologic characteristics. Based on the formation conditions of volcanic rock and the general mechanisms and diagenesis pattern of volcanism, volcanic facies is divided into volcanic conduit facies, explosive facies, effusive facies, extrusive facies, subvolcanic facies and volcanic sedimentary facies. There are explosive facies (volcanic breccia), effusive facies (basalt and andesite) and volcanic sedimentary facies (sedimentary tuff) in the study area.

The lithology of each drilled well in the study area is as follows. In Well HQ101, volcanic breccia (i.e., explosive facies) is dominant. There are fairly strong oil and gas shows in the interval of 1700–1800 m and discontinuous oil and gas shows in the interval of 1450–1650 m, indicating fairly good oil and gas shows on the whole in this well.

In Well HQ3, except tuff in the lower 150 m and the upper 40 m, the rest of the intervals are volcanic breccia, showing explosive facies is dominant in this well. There are excellent oil and gas shows in the interval of 2630–2720 m (volcanic explosive facies) and discontinuous oil and gas shows in the interval of 2780–2840 m (volcanic sedimentary facies) in this well, indicating that volcanic explosive facies is superior to volcanic sedimentary facies in reservoir quality. Well HQ102 has fairly good oil and gas shows on the whole, with the cycle of explosive facies→volcanic sedimentary facies transiting gradually to the one of effusive facies→volcanic sedimentary facies.

In Well HS1, the lithology from the bottom to the top is andesite→tuff→andesite→tuff, volcanic facies is effusive facies→volcanic sedimentary facies→effusive facies→volcanic sedimentary facies, and oil and gas shows get worse gradually with the increase of volcanic sedimentary facies upward.

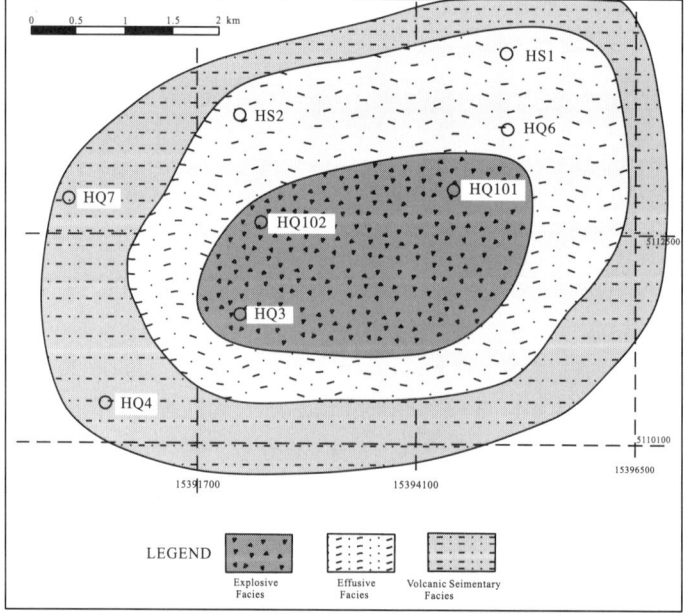

Figure 2-18 Plane distribution of volcanic rock lithofacies in the study area.

It is shown from the spatial superimposition relationship of volcanic facies in all wells of the study area that there are multiple volcanic eruption cycles there, in each cycle, there develops volcanic conduit facies, explosive facies, effusive facies and volcanic sedimentary facies in sequence with the increase of the distance from the crater.

It is shown in the plane distribution diagram of each well that volcanic explosive facies is mainly distributed in the area of Well HQ101 and HQ3, and to the northwest, volcanic eruption intensity weakens gradually, and volcanic facies changes in a trend of explosive facies→ effusive facies→volcanic sedimentary facies. It is inferred that volcanic breccia should also occur in the symmetry direction of the line from Well HQ101 to Well HQ3 (Figure 2-18).

Due to different formation environment, diagenesis and mineral composition, different volcanic facies have different reservoir physical properties, which are summarized in Table 2-2.

Table 2-2 Volcanic lithofacies classification and reservoir space types.

Lithofacies type	Lithologic characteristics	Location and diagenesis	Reservoir space types
Explosive facies	Tuff with vitreous fragment, crystal and magma clastics, and welded volcanic breccia	Product of volcanic explosion, closer to the crater; condensing cementation and compaction	Interbreccia pore, condensed shrinkage fracture, intracrystalline blast fracture, intracrystalline dissolved pore, matrix dissolved pore and structural fracture
Effusive facies	Vesicular basalt, amygdaloid basalt, andesite, vesicular rhyolite and autobreccia	The facies belt beyond explosive facies; condensing consolidation	Vesicle, intracrystalline dissolved pore, and dissolved pore in vesicle filler
Volcanic sedimentary facies	Laminated volcaniclastic rock and tuff	Distant end of volcano; compaction	Intramatrix dissolved pore and structural microfracture

1. Explosive facies

Volcanic explosive facies, often volcanic breccia is widespread in the study area, often accumulating near the carter. Composed of breccia and interbreccia filler, volcanic breccia contains dissolved pore, interbreccia pore and fracture etc. Mostly the product near the crater, volcanic breccia is formed at the early and climax stages of volcanic explosion when magma erupts violently. A rock framework is put up with volcanic breccia, inside fill fine grained materials (e.g., volcanic ash), and interbreccia pores are likely to form in the frameworks not completely filled with volcanic ash. The volcanic breccia has higher primary porosity, so various fluids can flow easily in it in the process of geological evolution, and dissolve the original minerals, especially the interbreccia matrix, as a result, volcanic breccia is often abundant in dissolved pores. Dissolution tends to happen again inside the dissolved pores, giving rise to multiphase mineral filling sequence of zeolite, quartz and calcite etc. Volcanic breccia is mainly generated when volcanic materials erupt to the air, fall down and accumulate, in the course, rock-forming materials condense quickly in the air, so volcanic breccia often has many primary blast fractures. In addition, as there is greater rigidity difference between volcanic breccia and interbreccia fillings (e.g., volcanic ash), secondary fractures tend to form inside volcanic breccia during the tectonic disturbance after the diagenesis.

With developed primary pores, primary fractures, secondary dissolved pores and secondary fractures, volcanic explosive facies (volcanic breccia) has higher porosity and fairly good connectivity between pores, and therefore, fairly good reservoir quality, and active oil and gas shows.

2. Effusive facies

There develop volcanic effusive facies in the study area, often basalt and andesite. Mostly forming at the middle stage of volcanic eruption cycle, after the violent volcanic explosion, effusive facies is formed by thicker magma erupting out of the strata due to the high pressure of the crust flowing on the surface under the joint driving of subsequent extrusion and its own gravity. In the process of flowing, the top, bottom and front margin of lava flow cool down more quickly, forming vesicular lava, so volcanic effusive facies has some primary vesicles. The magma inside lava flow cools down more slowly, so crystals are crystallized sufficiently and automorphic crystalline minerals often form. The crystalline minerals are in high degree of order, so the later fluids tend to dissolve the weak spots (e.g., cleavage fissure), producing intracrystalline dissolved pores. Effusive facies rocks (e.g., basalt and andesite) are also affected by tectonic stress during the later tectonic activities, and structural microfractures tend to form inside the rocks to compensate the rock stress, so there are secondary fractures in effusive facies.

Volcanic effusive facies (basalts and andesite) has primary pores, intracrystalline dissolved pores and secondary microfractures. Secondary microfractures, as important fluid migration and reservoir pathways, increase the flowing capacity of fluids and connects primary pores, so they are an important pore type in volcanic effusive facies. Volcanic effusive facies has some reservoir space, but poorer reservoir quality than volcanic explosive facies, and average oil and gas shows.

3. Volcanic sedimentary facies

There is volcanic sedimentary facies, tuff, in the study area. Formed at minor-gap stage of volcanic eruption, volcanic sedimentary facies is the accumulation of volcanic ash with grain size of less than 2 mm (e.g., lithic, crystal and vitreous fragment) far away from the crater after long-distance floating in the air. With fine grain size, larger surface area of particles, and unstable vitreous fragment, tuff tends to be reworked secondarily, as a result, there are intramatrix dissolved pores in volcanic sedimentary facies. Due to the later tectonic activities, tuff also has structural microfractures. The structural microfracture plays a decisive role in the multi-layer migration of fluid in tight tuff.

There are intramatrix dissolved pores and structural microfractures in volcanic sedimentary facies. The structural microfracture is the decisive factor for the development of reservoir space in volcanic sedimentary facies. Volcanic sedimentary facies has poorer reservoir quality than volcanic explosive facies and effusive facies, and poorer oil and gas shows.

The pore development characteristics of volcanic reservoirs in the study area show that the reservoir quality and oil and gas shows of drilling and mud logging get worse from volcanic explosive facies to effusive facies to sedimentary facies.

2.3.2 Effect of dissolution

Volcanic rocks are derived from cooling of magma. After diagenesis, the minerals in pore fluid precipitate and crystallize in stable environment, giving rise to quartz, calcite, chlorite and zeolite, which fill primary pores and fractures, lowering reservoir quality. It is shown in Figure 2-19a and b that there are automorphic calcite crystals in the volcanic breccia, and the primary component between volcanic breccias was originally volcanic ash, after stage-one dissolution, automorphic calcite replaced the volcanic ash, after secondary dissolution and precipitation, stage-two calcite was formed, then, the stage-two calcite was cut by the stage-three calcite; and stage-four dissolution happened between stage-four calcite crystals, forming intracrystalline pores filled with crude oil. Initially, the andesite and tuff are structurally tight. It is shown in Figure 2-19c and d that there are dissolved pores and dissolved fractures in tight tuff and the dissolved fractures are filled with dark bitumen.

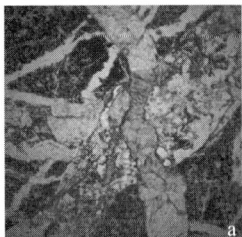

Multiphase calcite filling, Well HQ6, 140.0 m plane-polarized light.

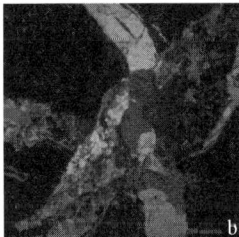

Multiphase calcite filling, Well HQ6, 140.0 mcross-polarized light.

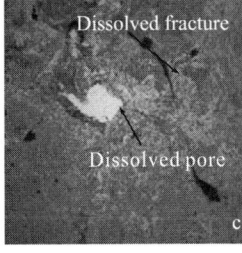

Dissolved pores and fractures, WellHQ6, 300.9 m, plane-polarized light.

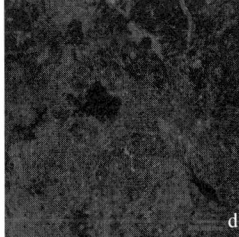

Dissolved pores and fractures, Well HQ6, 300.9m, cross-polarized light.

Figure 2-19 Multiphase dissolution and filling of fractures in volcanic reservoirs in the study area.

Dissolution is a kind of important diagenetic epigenesis in the study area, which can increase reservoir space and permeability of the rock. The filler in all reservoir space is mostly dissolved again by the later fluids, giving birth to intracrystalline dissolved pore, intercrystalline dissolved pore, intrabreccia dissolved pore and interbreccia pore, and oil and gas shows often occur in the dissolved pores.

2.3.3 Effect of tectonic activity

Structural faults are the pathways for magma to flow upward, and tectonic activities control volcanic eruption and planar distribution of volcanic rock. In addition, fractures generated by tectonism are favorable flowing pathways of formation fluid in volcanic massif, allowing materials to exchange smoothly between formation fluid and volcanic massif, and providing the basis for the occurrence of secondary diagenesis. Therefore, volcanic rock with developed structural fractures also has stronger secondary diagenesis. Multiphase diagenesis can be often identified in the structural fractures preserved in the volcanic rock in the study area, and it is mainly presented as the superimposition of filling-dissolution of different periods.

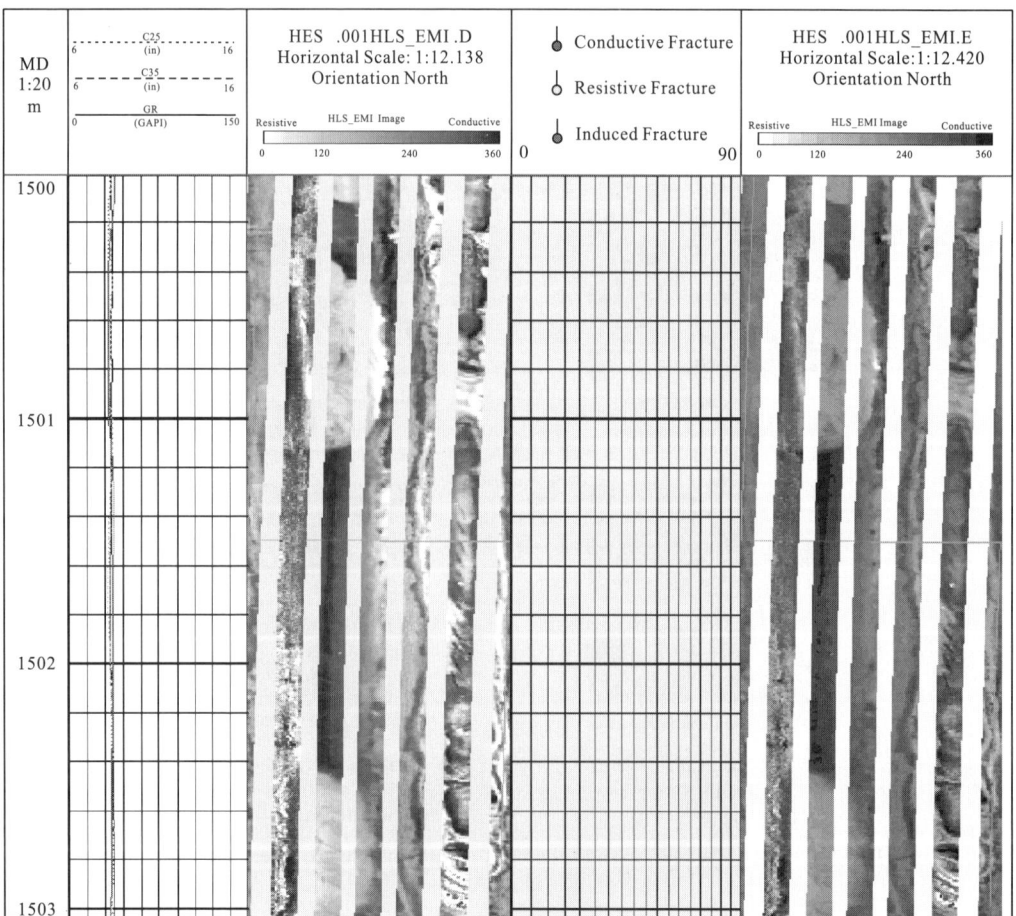

Figure 2-20 XRMI imaging characteristics of dissolved pores in Well HQ6 (1150–1153 m).

Going through complicate structural evolution and multiple phases of tectonic activities, the study area has complex faults and a great number of induced fracture systems. Wells drilled in the study area reveal that the Carboniferous strata is mainly composed of volcanic rocks, including andesite, tuff, basalt and volcanic breccia; and the Permian strata is mainly composed of mudstone, sandy mudstone and andesite. The development characteristics of fractures in the Carboniferous are obviously different form those in the Permian because of lithologic difference and complicate structural evolution process. At macro level, it is shown from XRMI dissolved pore imaging and some drilling cores that high-angle fractures (70°–80°) and low-angle oblique fracture (30°–45°) are developed in the massif under the action of tectonism, the imaging color at the wall along the fractures is dark black, indicating that the fractures act as high-conductivity layers, are mostly filled with low-resistivity fluids (e.g., oil and gas). To sum up, fractures are good fluid migration pathways and reservoirs space (Figure 2-20).

At micro level, it is found from observation of cores and thin sections that most of the fractures cut through primary pores and secondary dissolved pores in the volcanic rock that are originally independent, disconnected or poorly connected with each other, improving the porosity and permeability of the volcanic rock (Figure 2-21). In addition, fractures are the pathways for deep fluids flowing upward to the shallow formations, and the fluids can dissolve pore filler, thus enlarging fracture width and improving reservoir quality. In this way, a virtuous cycle is built up between fractures and fluids.

Structural fractures in dioritic porphyrite, Well HS1, 2553.00 m.

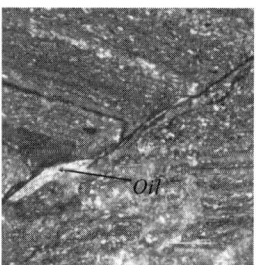

Structural fractures in oil-patched mudstone, Well HS1, 2094.70 m, cross-polarized light.

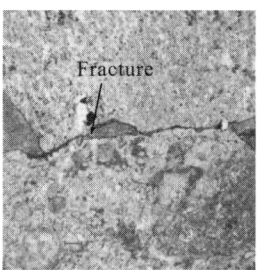

Oil-gas bearing fractures connected primarypores, Well HQ6, 814.10 m, plane-polarizedlight.

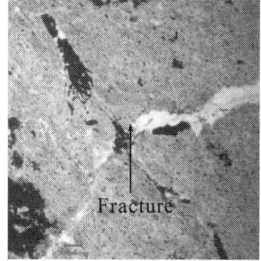

Oil-gas bearing fractures connected primarypores, Well HQ6, 814.10 m, cross-polarized light.

Figure 2-21　Reworking effect of fractures on Carboniferous volcanic reservoir.

Chapter 3

Oil-source Correlation and Migration Pathway Tracing

Igneous rocks cannot generate oil and gas themselves, thus the primary condition for forming igneous reservoirs is necessary oil source condition, therefore, igneous rock must be associated with source rock or near source sags to form large oil and gas reservoirs. There are several sets of source rocks, including Carboniferous, Permian, Triassic and Jurassic systems, etc. in the northwest margin of Junggar Basin(Ding et al., 1994), among which, Permian system is the major source rock series. Up to now, Xinjiang Oilfield Company has discovered several oil and gas fields sourced by Permian system at the northwest margin of the Junggar Basin(Zhang, 2013).

Situated in the overthrust tectonic belt of the Hala'alat Mountain and north of the Mahu Sag, Junggar Basin, the Hala'alate Mountain area is in the direction of migration and accumulation of hydrocarbons generated in the hydrocarbon generation center of the Mahu Sag(Zhang, 2013; Li, 2013), and is an important part of the Mahu-rim oil and gas system. During early exploration stage in this area, Sinopec Shengli Oilfield mainly explored the onlap-denudation belt in shallow strata, discovered Chunhui Oilfield and Arad Oilfield, and proposed meshwork-carpet type hydrocarbon accumulation theory(Wang et al., 2014). As Well HQ6 and Well HS1 encountered dark mudstone in Permian Fengcheng Fm in deep strata of the thrust nappe, obtaining low-production and medium oil flow by fracturing, and thick intervals in Carboniferous volcanic rock with oil and gas shows were found in the thrust nappe, more and more researchers began to study the contribution of Permian source rock beneath the thrust nappe to hydrocarbon accumulation, and hydrocarbon accumulation mode in the thrust nappe (Wang et al., 2014; Zhang, 2013; Hu and Xia, 2012), thus accurate oil-source correlation is particularly important.

In this chapter, we systematically analyze and compare the geochemical characteristics of source rocks and crude oil in various tectonic belts in this study area, then clarify the source of hydrocarbons in various layer series. We will find out the hydrocarbon conducting system in the study area by tracing hydrocarbon migration pathways, so as to secure a strong basis for establishing hydrocarbon accumulation model and analyzing hydrocarbon distribution pattern, and provide guidance and reference for exploring hydrocarbons and predicting favourable zones in the Hala'alate Mountain area.

3.1 Geochemistry of source rocks

3.1.1 Evaluation criterion of source rocks

Organic matter abundance, one of the major parameters for evaluating hydrocarbon generation capacity of source rock, reflects the quantity of hydrocarbon generation matters in source rock. Generally, quantitative estimation of organic matter abundance of source rock is to measure organic carbon content, chloroform bitumen "A", total hydrocarbon content and rock pyrolysis parameter S_1+S_2. A lot of research on continental source rock has been made in China, and some standard evaluation criteria for continental source rock have been established, among which, the evaluation criterion proposed by Qin Jianzhong is the most widely used in China, hence, this evaluation criterion was adopted to evaluate organic matter abundance of the continental source rocks in the Hala'alate Mountain area in this research (Table 3-1).

Organic matter type evaluation is another important task in evaluating hydrocarbon generation capacity of source rock. Different types of source rock are different in hydrocarbon generation potential, hydrocarbon type generated and threshold temperature. Organic matter types in source rock can be evaluated by analyzing soluble organic matter (bitumen) and insoluble organic matter (kerogen). The commonly used methods include element analysis, optic analysis, infrared spectrum analysis and rock pyrolysis analysis etc.

Table 3-1 Evaluation criterion for organic matter abundance of muddy source rock in continental lakes of China.

Evolution stage	Source rock grade / Evaluation parameters	Rich or very good	Good	Moderate	Poor	None
Immature-mature	TOC/%	>2.0 (I - II$_1$) >4.0 (II$_2$-III)	1.0-2.0 (I - II$_1$) 2.5-4.0 (II$_2$-III)	0.5-1.0 (I - II$_1$) 1.0-2.5 (II$_2$-III)	0.3-0.5 (I - II$_1$) 0.5-1.0 (II$_2$-III)	<0.3 (I - II$_1$) <0.5 (II$_2$-III)
	Chloroform bitumen "A" /%	>0.25	0.15-0.25	0.05-0.15	0.03-0.05	<0.0.3
	HC/(μg/g)	>1000	500-1000	1.50-500	50-150	<50
	S_1+S_2/(mg/g)	>10.0	5.0~10.0	2.0~5.0	0.5~2.0	<0.5
High mature-over mature	TOC/%	>1.2 (I - II$_1$) >3.0 (II$_2$-III)	0.8-1.2 (I - II$_1$) 1.5-3.0 (II$_2$-III)	0.40-0.8 (I - II$_1$) 0.6-1.5 (II$_2$-III)	0.2v0.40 (I - II$_1$) 0.35-0.6 (II$_2$-III)	<0.20 (I - II$_1$) <0.35 (II$_2$-III)

Organic matter maturity is one of the important indexes for evaluating hydrocarbon generation capacity of source rock, and an important basis for evaluating hydrocarbon generation quantity and resource potential of a region or a set of source rock. In recent years, many methods and indexes have been proposed for this issue in China and other countries. The commonly used maturity indexes include R_o, T_{max}, $S_1/(S_1+S_2)$, H/C atom ratio of kerogen, soluble organic matter content in rock and biomarkers etc. Among them, R_o has the best effect, and T_{max} is the second in effect. Other parameters are usually used as assisting indexes (Table 3-2).

Table 3-2 T_{max}(℃) scope of source rock in China.

Thermal evolution stage		Immature	Oil	Gas condensate	Wet gas	Dry gas
	R_o/%	<0.5	0.5-1.3	1.0-1.5	1.3-2	>2
T_{max}/℃	Type I organic matter	<437	437-460	450-465	460-490	>490
	Type II organic matter	<435	435-455	447-460	455-490	>490
	Type III organic matter	<432	432-460	445-470	460-505	>505

3.1.2 Evaluation of source rocks

As the formations in the Hala'alate Mountain area are shallower, we only obtained source rock samples from Permian system in the Hala'alate Mountain area, the source rock samples of Jurassic and Triassic systems were obtained from Well Y1 in adjacent Shixi Sag. The total depth of Well Y1 is 2800 m. The source rock samples for geochemical analysis taken from this well cover all the formations from Cretaceous Tugulu Group to Upper Triassic Baijiantan Fm, which provide plentiful initial data for evaluating source rock.

1. Jurassic source rock

According to the drilling data of Well Y1, dark mudstone is better developed in Lower Jurassic Badaowan Fm and Sangonghe Fm, and Middle Jurassic Xishanyao Fm, accounting for more than 40% each; little dark mudstone occurs in Middle Jurassic Toutunhe Fm. Coal layers are mainly distributed in Lower Jurassic Badaowan Fm and Middle Jurassic Xishanyao Fm, and thicker in Xishanyao Fm (with a total thickness of 11 m), but thinner in Badaowan Fm (with a total thickness of 5 m) (Figure 3-1).

Dark mudstone in Jurassic Xishanyao Fm and Sangonghe Fm in Well Y1 has a TOC of about 0.5% in general, and pyrolysishydrocarbon potential (S_1+S_2) of less than 2 mg/g. Dark mudstone of 1800-1844 m section in Jurassic Xishanyao Fm has an average TOC of 0.49%, and an average S_1+S_2 of 1.78 mg/g; dark mudstone of 1896-2007 m section in Sangonghe Fm has an average TOC of 0.42%, and average S_1+S_2 of 1.21 mg/g, which show these two sections of dark mudstone are poor source rock, and the others are not source rock. Dark mudstone of 2318-2423 m section in Jurassic Badaowan Fm has a TOC of more than 0.5% in general

(3.47% at maximum), and S_1+S_2 between 2–3.5 mg/g, showing its organic matter abundance reaches the standard of better source rock, and its organic matter is mainly type Ⅱ-Ⅲ.

Coal seam and dark mudstone in Jurassic Xishanyao Fm and Badaowan Fm in Well Y1 have a R_o between 0.48%–0.585% (generally around 0.5%), indicating that all the organic matter in Jurassic dark mudstone in this well is at immature-low mature stage, and the source rocks have not reached hydrocarbon generation peak. As Well Y1 is at the bulge of the basin margin, the Jurassic dark mudstone there has a smaller absolute thickness than that in the basin centre (the max thickness of the dark mudstone in Xishanyao Badaowan Fm is about 400 m in the Mahu Sag), it is inferred that as the center of the Mahu Sag had better sedimentary environment, the source rocks would have bigger thickness and higher organic matter abundance. Correspondingly, with the increase of buried depth, the conversion of sedimentary organic matter to hydrocarbons is more favourable.

2. Triassic source rock

Dark mudstone of 2519–2571 m thick in Upper Triassic Baijiantan Fm in Well Y1 has type Ⅱ-Ⅲ organic matter, a TOC of more than 0.5% (1.57% at most), and S_1+S_2 between 2–3.5 mg/g (5.4 mg/g at maximum), reaching the standard of poor source rock.

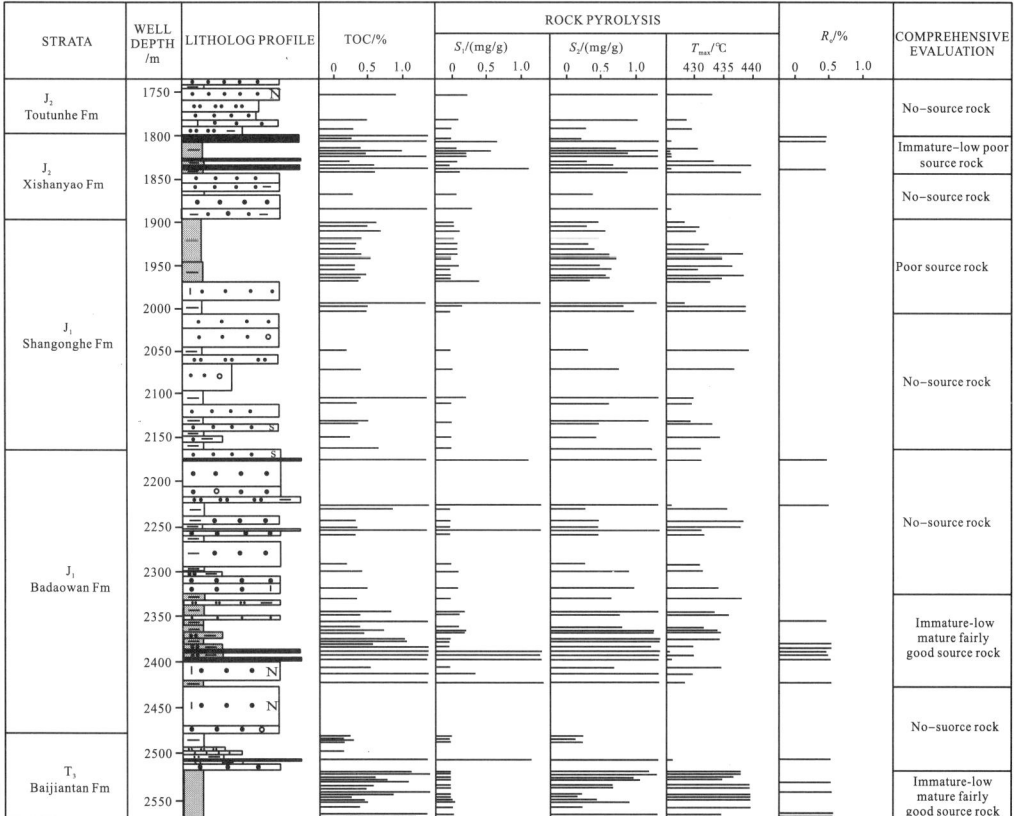

Figure 3-1 Geochemical profile of Jurassic-Triassic source rock in Well Y1.

In Well Y1, out of seven dark mudstone samples taken from Upper Triassic Baijiantan Fm, three samples (2486.28 m, 2487.58 m and 2489.25 m) have low organic matter abundance with TOC between 0.18%–0.34%, indicating they are not source rock; while four analysed cutting samples between 2529–2571 m section have a TOC of 1.08%–2.72% (on average 1.64%), reaching good source rock standard, which is consistent with the high organic matter abundance of the section shown by geochemical logging (Figure 3-1).

From the pointview of maturity, Upper Triassic Baijiantan Fm in Well Y1, with a R_o of 0.54%–0.58%, is at low mature stage. According to results of previous research, Triassic Baijiantan Fm has generated and expelled a smaller quantity of oil, which can hardly form industrial oil accumulation in the Hala'alate Mountain area after long distance migration(Wang et al., 1989).

3. Source rock in Permian Fengcheng Fm

The Permian Fengcheng Fm (P_1f) has been proved an important set of source rock in the north-western margin of the Junggar Basin. In the Mahu Sag, it is thick and widely distributed, and is mainly dark-grey mudstone; with an area 6340 km^2, max deposition thickness of about 1800 m, it is about 7608 km^3 in volume. The source rock in Lower Permian Fengcheng Fm in the Mahu Sag has an average TOC of 1.26%, chloroform bitumen "A" of 1493 ppm, total hydrocarbon content of 820 ppm, and S_1+S_2 of 7.3 mg/g. The source rock has mainly type I kerogen, a R_o of 0.85%–1.16%, indicating the source rock is at mature stage, and is a set of fairly good -good source rock.

The analysis results of source rock samples from the Fengcheng Fm in the Hala'alate Mountain area show the source rock has a TOC of 0.29%–5.35% (on average 1.35%), chloroform bitumen "A" of 0.0178%–0.7525% (on average 0.226%), and S_1+S_2 of 1.29–17.7 mg/g (on average 5.60 mg/g), therefore, it can be seen that the source rock of Lower Permian Fengcheng Fm in the Hala'alate Mountain area has higher organic matter abundance, representing good source rock.

In the macerals, the source rock has main sapropelite, developed exinite and very low inertinite content (Table 3-3); an average H/C ratio in kerogen of 1.17, lighter carbon isotope and chloroform bitumen "A" ($\delta^{13}C$ of -28.72‰ to -31.95‰, on average -30.0‰), and mainly type I - II$_1$ kerogen.

Table 3-3 Maceral identification and type division of kerogen in Fengcheng mud of Well HQ6.

Interval/m	Lithology	Sapropelite /%	Exinite/%	Vitrinite/%	Inertinite /%	Type	Type index
1410–1425	Grey mud	88	0.7	10	1.3	II$_1$	79.5
1450–1460	Grey mud	83	0	14	3	II$_1$	69.5
1494–1502	Grey mud	92.3	1.7	5.3	0.7	I	88.5
1560–1600	Grey mud	92.3	1.3	5.7	0.7	I	88.1
1600–1630	Grey mud	86	3.3	9.7	1	II$_1$	79.4
1630–1660	Grey mud	89.7	2.3	7.3	0.7	I	84.7

The source rock in drilled wells in the Mahu Sag has a measured R_o of 0.75%–2.02%, and source rock in Well HQ6 etc. in the Hala'alate Mountain overthrust belt has a measured R_o of 0.75%–0.94% (on average 0.85%), indicating mature oil generation stage.

4. Source rock in Permian Wuerhe Fm

The Wuerhe Fm (P_2w), mainly dark mudstone and silty mudstone, is another set of good source rock in the Junggar Basin. The analysis results of rock samples show that this set of source rock has a TOC of 0.5%–1.5% (on average 1.01%), average chloroform bitumen "A" of 0.283%, and average S_1+S_2 of 0.28 mg/g; $\delta^{13}C$ in chloroform bitumen "A" of −29.5‰ to −26.4‰ (on average −27.9‰), and $\delta^{13}C$ in kerogen of −22.0‰ to −20.0‰ (on average −21.1‰), and mainly type III and type II$_2$ organic matter. This set of source rock in the Mahu Sag has a measured R_o of 0.51%–1.86%, indicating it is at mature-high mature stage.

5. Source rock in Carboniferous

In terms of drilling data, influenced by overthrust tectonic belt in this study area, some wells (HQ6, HQ4, HS1, etc.) all encountered residual Carboniferous formation, revealing its main lithology is brown and grey tuff and volcanic breccia, with no source rock found.

Carboniferous outcrops are widely distributed in Hala'alate Mountain profile, Burqin South profile, Tost Southeast profile and Baiyangzhen North profile etc. (Figure 3-2). Carboniferous mudstone on these profiles is mainly grey-greyish black silty mudstone.

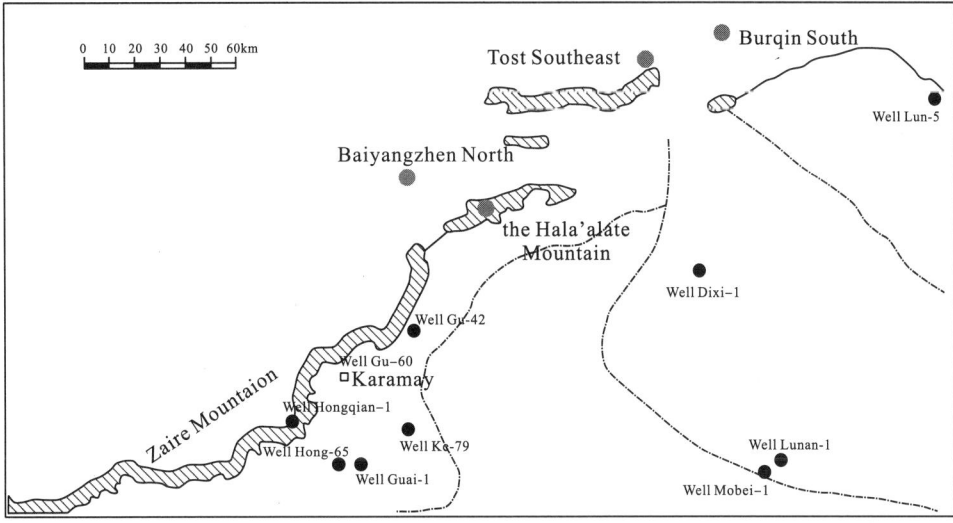

Figure 3-2 Carboniferous outcrop locations in Hala'alate Mountain area.

The the Hala'alate Mountain profile is situated at about 20 km northeast of Wuerhe along national road No.201 (Figure 3-3). On this profile, the upper part is the Aladeyikesai Fm grey-greyish black tuffaceous sandstone, tuffaceous siltstone and fine sandstone, and greyish black silty mudstone, with a thickness of 1300 m; the lower part is the Hala'alate Fm greyish green-grey volcanic breccia and tuffaceous breccia, and greyish green andesite and andesitic basalt. The major source rock interval is in the upper Aladeyikesai Fm, with a thickness of up to 700–800 m. As the source rock has poorer weathering resistance, the valley with lower terrain was formed; influenced by long-term current scouring, the source rock on the outcrop is in lamellar shape. The Hala'alate Fm mainly consists of greyish green and grey volcanic rock, with no mudstone.

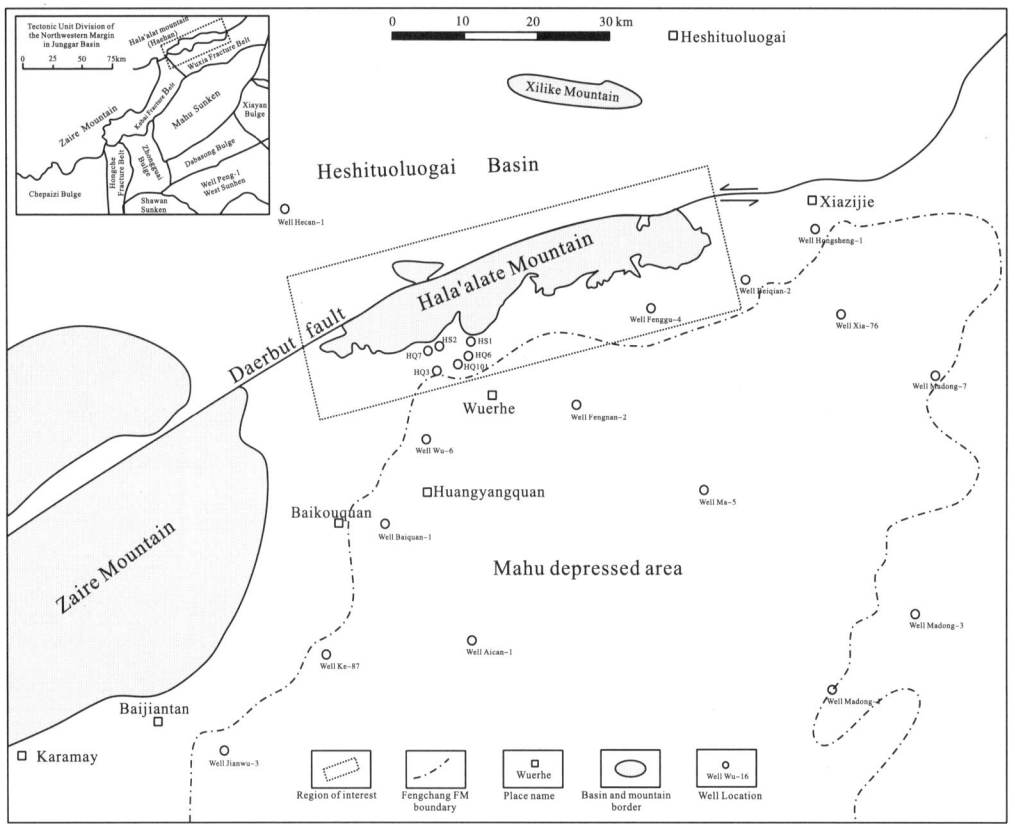

Figure 3-3 Location of the Hala'alate Mountain profile.

The outcrop samples on the Hala'alate Mountain profile and the middle Hala'alate Mountain profile are black silty mudstone with lower organic matter abundance. They have a TOC of 0.18%–0.71% (on average 0.41%), chloroform bitumen "A" of 0.0035%–0.0082%, "S_1+S_2" of 0.01–0.05 mg/g, and T_{max} of 424–501 ℃. The outcrop samples from the middle Hala'alate Mountain profile have higher organic matter abundance: with a TOC of 0.11%–1.44% (on average 0.88%), chloroform bitumen "A" of 0.0016%–0.0030%, "S_1+S_2" of 0.01–0.05 mg/g, and T_{max} of 416–543 ℃. In summary, the Hala'alate Mountain profile has poorer organic matter abundance (Table 3-4).

The outcrop samples from the Baiyangzhen profile are mainly black silty mudstone with lower organic matter abundance. They have a TOC of 0.18%–0.61% (on average 0.4%), chloroform bitumen "A" of 0.0019%–0.0045%, "S_1+S_2" of 0.01–0.03 mg/g (on average 0.02 mg/g), and T_{max} of 399–494 ℃. In summary, the Baiyangzhen profile has poorer organic matter abundance (Table 3-4).

Table 3-4 Geochemical data of Carboniferous profiles in the Hala'alate Mountain area.

Profile	Strata	Lithology	Range	TOC/%	S_1+S_2/(mg/g)	T_{max}/℃	Organic matter type	R_o/%
the Hala'alate Mountain	C_2a	Greyish black silty mudstone	Value interval	0.18–0.71	0.01–0.05	424–501	II_1, III	2.56–3.06
			Average	0.41	0.028	463		2.79
the Middle Hala'alate Mountain	C_2a	Greyish black silty mudstone	Value interval	0.17–1.44	0.02–0.08	416–543	II_1, III	1.22–1.78
			Average	0.88	0.038	490		1.46
Baiyangzhen North	C_1n	Grey black silty mudstone	Value interval	0.18–0.61	0.01–0.03	399–494	I, II_1	0.92–2.35
			Average	0.4	0.02	440		1.53
Tost Southeast	C_1h	Coal	Value interval	0.25–15.2	0.01–0.17	498–543	I, III	1.69–3.20
			Average	5.79	0.08	519		2.16
Burqin South	C_2q	Greyish black silty mudstone	Value interval	0.27–1.21	0.02–0.04	464–542	II_1, III	1.36–1.42
			Average	0.81	0.03	500		1.39

The outcrop samples from the Tost Southeast profile are coal and black silty mudstone, they have a TOC of 0.25%–15.2% (on average 5.79%), chloroform bitumen "A" of 0.0022%–0.0508% (average 0.023%), "S_1+S_2" of 0.01–0.17 mg/g (on average 0.08 mg/g), and T_{max} of 498–543 ℃, indicating fairly high organic matter abundance (Table 3-4).

The Carboniferous outcrop samples from the Burqin South profile are black silty mudstone. They have a TOC of 0.27%–1.21% (on average 0.81%), chloroform bitumen "A" of 0.0014%–0.0023%, "S_1+S_2" of 0.02–0.04 mg/g (on average 0.023 mg/g), and T_{max} of 462–542 ℃, indicating lower organic matter abundance (Table 3-4).

In terms of organic matter type, the Carboniferous source rock in the Hala'alate Mountain area shows diversity (including type I, type II and type III), and is at high-over mature stage (Table 3-4). According to the analysis results of inclusions in surface rocks of the Hala'alate Mountain area, and sterane-terpane biomarker data in its surrounding regions (Chen and Kuang, 2010; He et al., 2010), during Permian Period, the Carboniferous source rock should be at the stage of peak oil generation, and hydrocarbon generation, migration and accumulation had occurred; however, as the tectonic activity during this period was quite violent, and hydrocarbon accumulation and preservation conditions were poorer, thus hydrocarbons were hard to accumulate. Moreover, there is a big maturity difference between Carboniferous and Permian source rocks, which means that the Carboniferous source rock had reached high-over mature stage before the deposition of Permian system, which is unfavourable for the accumulation of hydrocarbons generated by Carboniferous system.

3.1.3 Biomarkers of source rocks

1. Source rock in Jurassic system

The normal paraffin hydrocarbons in coal-measure mudstone of Jurassic Badaowan Fm have carbon number distribution scope of nC_{11}-nC_{32}, and a larger peak carbon number (mainly nC_{23} and nC_{25}), with apparent odd/even predominance (with OEP generally >1.50), indicating the major hydrocarbon generation materials are terrestrial higher plants, with Pr/Ph between 3-11, lack of carotane, which is related to the oxidation prone nature of the environment during early coal forming (Chen and Kuang, 2010; He et al., 2010; Wang et al., 1989).

The Jurassic source rock has lower tricyclic terpane abundance, carbon numbers in decrease distribution of C_{20}>C_{21}>C_{23}, Ts relative abundance much lower than Tm relative abundance, and lower gammacerane content, indicating the depositional water body had lower salinity. The source rock has fingerprint of regular sterane $\alpha\alpha\alpha20RC_{27}$, C_{28} and C_{29} of inverse "L" shape or "V" shape; slightly dominant $\alpha\alpha\alpha20RC_{29}$ sterane, indicating the main organic matter input is terrestrial higher plants (Figure 3-4).

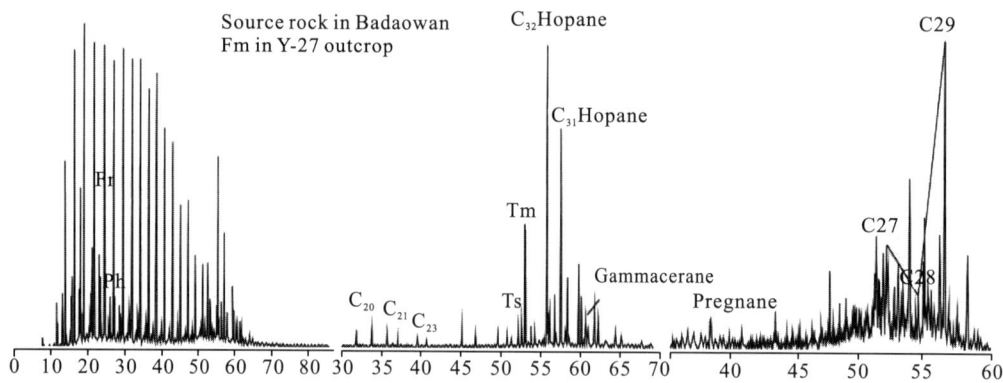

Figure 3-4 Biomarker spectrum of source rock in Jurassic Badaowan Fm.

2. Source rock in Permian Fengcheng Fm

Permian Fengcheng Fm source rock has fairly complete normal paraffin hydrocarbons (Figure 3-5), higher content of isoprenoid paraffin hydrocarbons (pristine and phytane, etc.), the feature of predominant phytane (with Pr/Ph of 0.27-1.28, on average 0.60), which indicates that the source rock was deposited in reducing environment with high salinity and deficient oxygen. Its higher β-carotane content means that the organic matter input was related to pigment source of photosynthetic bacteria which were mainly generated in reducing environment(Tian et al., 2011; Peters and Moldowan, 1993; Jiang and Fowler, 1986), and the deposition speed was higher, favourable for organic matter preservation.

The source rock in the Fengcheng Fm has the following terpane and sterane biomarker features: higher gammacerane content, with a gammacerane index (gammacerane/C_{30} hopane ratio) of 0.11–0.64 (on average 0.31), reflecting brackish-saline depositional environment. The source rock has higher tricyclic diterpane content, increase tendency of C_{20}, C_{21} and C_{23}; trace or undetectable Ts content (Figure 3-1); regular sterane content in increasing distribution ($\alpha\alpha\alpha20RC_{27} \ll \alpha\alpha\alpha20RC_{28} < \alpha\alpha\alpha20RC_{29}$), with content of 3.2%–20.9%, 30.7%–43.7% and 42.7%–56.2%, respectively, which indicate that the organic matter is from low aquatic organism (such as fungi and algae). Source rock samples from Well HQ6 etc have a Boron content of 68.9–202.4 ppm, and Sr/Ba ratio of 0.93–1.45, indicating that the source rock was deposited in strong reducing saline lacustrine environment. In summary, the source rock in the Fengcheng Fm is a set of fairly good-good source rock.

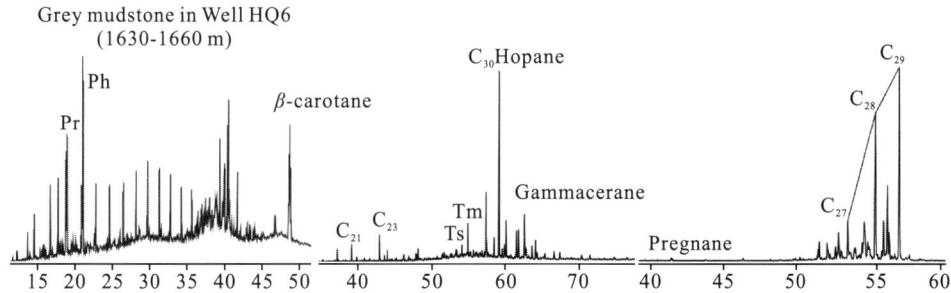

Figure 3-5 Biomarker spectrum of source rock in Fengcheng Fm of Well HQ6.

3. Source rock in Permian Wuerhe Fm

The source rock in the Wuerhe Fm has lower isoprenoid paraffin hydrocarbon abundance on gas chromatography, an average Pr/Ph value of 1.23 (pristane in dominance), and rare β-carotane (Figure 3-6).

The spectrum of sterane and terpene shows the source rock in the Wuerhe Fm has C_{23} as the main peak of tricyclic diterpane, lower content of $18\alpha(H)$-22, 29, 30-trisnorhopane (Ts), lower content of gammacerane, abundance of regular sterane C_{27}, C_{28} and C_{29} in increase tendency; $\alpha C_{27}/C_{29}$ ratio of $\alpha\alpha\alpha20R$ sterane of 0.3–0.8, and C_{28}/C_{29} ratio of $\alpha\alpha\alpha20R$ sterane of 0.3–0.9. The analysis results of Well HSX1 trace elements show the Wuerhe Fm source rock has a boron element content of 16.52–46.05 ppm, and a Sr/Ba ratio of 0.69–0.96, indicating the source rock was formed in fresh water weak oxidation and weak reduction lacustrine depositional environment.

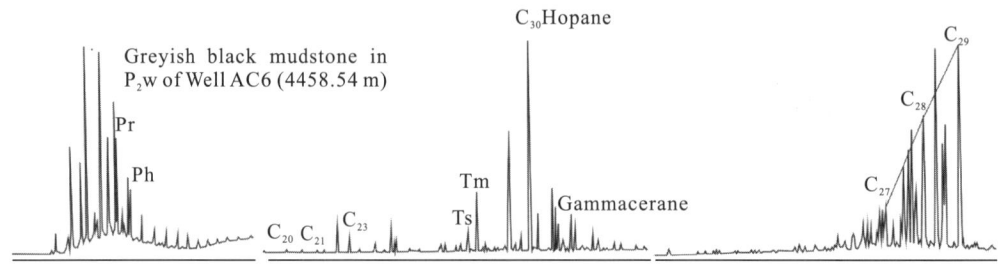

Figure 3-6 Spectrum of biomarkers in Wuerhe Fm source rock.

4. Source rock in Carboniferous system

The biomarker features of the source rock in Carboniferous system can be subdivided into the following two types:

The biomarker features of type I source rock are similar to that of the Permian source rock, with abundant β-carotane, Ts<Tm, higher tricyclic terpane and gammacerane contents, higher abundance of C_{28} and C_{29} in regular sterane $\alpha\alpha\alpha20RC_{27}$, $\alpha\alpha\alpha20RC_{28}$ and $\alpha\alpha\alpha20RC_{29}$, C_{28} slight lower than C_{29}, and very low relative abundance of C_{27} (Figure 3-7a).

The type II source rock has no β-carotane compound series in saturated hydrocarbons, and low carbon number hydrocarbons less than C_{23} taking the majority in normal paraffinic hydrocarbons. It has Pr/Ph of 0.41–0.63, with apparent pristane prevalence, abundant long chain tricyclic terpane, bigger variation in gammacerane abundance, a gammacerane/C_{30} hopane ratio of 0.17–0.50, indicating that water salinity was constantly changing. Its regular sterane $\alpha\alpha\alpha20RC_{27}$, $\alpha\alpha\alpha20RC_{28}$ and $\alpha\alpha\alpha20RC_{29}$ shows "V" shape distribution, C_{29}> C_{27}>C_{28}, indicating the original source materialsare mixed type organic matter(Tian et al., 2011); there is some 4-methyl sterane, and widely distributed dinosterane (Figure 3-7b).

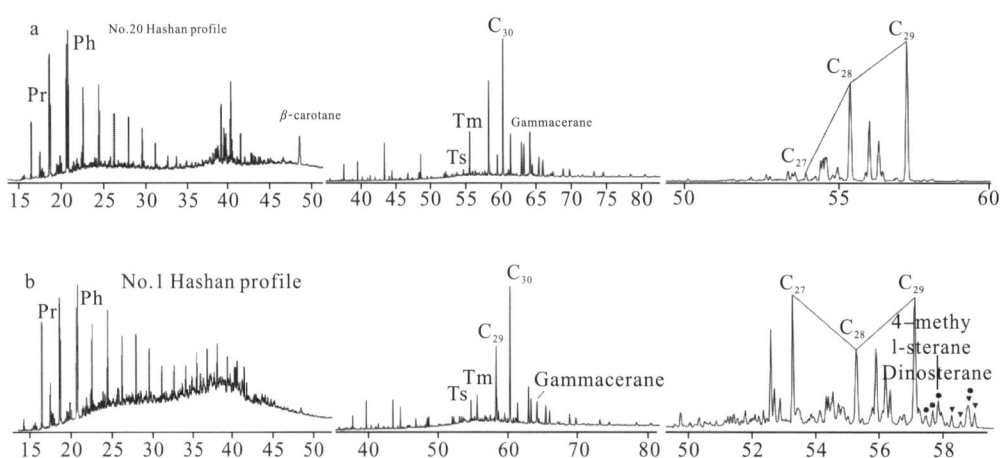

Figure 3-7 Spectrum of biomarkers in two types of Carboniferous source rock.

3.2 Geochemical features of crude oil and oil-source correlation

3.2.1 Physical properties of crude oil

Physical properties of crude oil are the macroscopic reflection of their chemical composition. For the Hala'alate Mountain area, crude oil of different layers have bigger differences in physical properties. According to density and viscosity etc. of crude oil, there are four types (medium oil, ordinary heavy oil, super heavy oil, extra heavy oil) of crude oil in the study area.

In the shallow onlap-denudation belt, the crude oil at 100–600 m burial depths is mainly super heavy oil and extra heavy oil. They have features of "four highs and two lows": high density, high viscosity, high freezing point, low wax content, and low sulfur content (Table 3-5). As the crude oil has suffered biodegradation and oxidization, their acid value is 0.39–7.84 mgKOH/g (on average 3.45 mgKOH/g), belonging to high acid crude oil.

In thrust nappe belt, the crude oil, at burial depth of 500–2800 m, has better quality (mainly ordinary heavy oil and medium oil), hereinto, the crude oil in the Fengcheng Fm belongs to medium oil with low sulfur content and low freezing point (Table 3-5). On the whole, different types of crude oil have different wax contents: the biodegraded oil in shallow layers has low wax content, while the normal crude oil in deep layers has higher wax content. On the plane, crude oil density and viscosity have feature of "lower in north and west, higher in south and east"; in vertical direction, with the decrease of burial depth, the crude oil turns poorer in physical properties, with increasing density, viscosity and freezing point.

Table 3-5 Physical properties of crude oil in various geologic units in Hala'alate Mountain area.

Geologic unit	Density at 20℃ /(g/cm^3)	Viscosity at 80℃ /(mPa·s)	Sulfur content /%	Wax content /%	Freezing point /℃
Onlap-denudation belt K, J, T	0.9535–1.0104	3919–68485	0.15–0.51	0.82–3.4	12–51
	0.9769(37)	10995(37)	0.38(5)	1.36(13)	38(25)
Thrust nappe belt P, C	0.8927–0.9139	58.7–476	0.24–0.38	3.27–10	−2 to −12
	0.9048(6)	263.3(6)	0.28(5)	6.59(4)	−6.4(5)

Note: the figure above the line is value interval, the figure beneath the line is average value and sample number.

3.2.2 Geochemical features of crude oil and oil-source analysis

1. Group composition features

Group composition of crude oil is a comprehensive reflection of its source, migration, dissipation and preservation conditions, etc. When all these conditions are similar, group composition of crude oil can reflect its source to a certain degree: the crude oil from the same source should have similar group composition, while the crude oil from different sources should have different group composition.

Group composition of crude oil in the Hala'alate Mountain area features "three lows and two highs": low saturated hydrocarbon/aromatic hydrocarbon ratio (0.947-4.027), low saturated hydrocarbon content, low aromatic hydrocarbon content, high non-hydrocarbon content and high asphaltene content. Density and viscosity of crude oil have better positive relationship with non-hydrocarbon and asphaltene contents. Overall, with the decrease of burial depth of oil reservoirs, saturated hydrocarbon content shows a decline trend, while the relative content of the non-hydrocarbons and asphaltene show an increase trend. In the crude oil of the Hala'alate Mountain area, the saturated hydrocarbons have $\delta^{13}C$ of -28.1‰ to -32.3‰ (on average -30.28‰), the aromatic hydrocarbons have $\delta^{13}C$ of -28.1‰ to -31.0‰ (on average -29.25‰), the non-hydrocarbons have $\delta^{13}C$ of -27.4‰ to -30.4‰ (on average -28.94‰), and the asphaltene has $\delta^{13}C$ of -27.4‰ to -29.8‰ (on average -28.65‰), which reflect that its hydrocarbon generation materials are from low aquatic organism (such as fungi and algae), with features of humic type of hydrocarbon generation materials.

2. Biomarkers of crude oil

1) Jurassic crude oil

Jurassic crude oil suffered serious biodegradation, thus has lost most normal paraffin hydrocarbons. Based on the differences of biomarker features of terpane and sterane, the Jurassic crude oil can be divided into two types. Oil-source correlation results indicate that they are from different source rocks.

Type I crude oil has higher sterane/hopane ratio (0.83-1.34), lower Ts/Tm ratio (<0.30), increasing distribution of C_{19}, C_{20} and C_{23} tricyclic terpane, little rearranged sterane and 4-methyl-sterane, and abundant gammacerane (gammacerane/C_{30} hopane >0.40), indicating brackish-saline depositional environment. Regular sterane $\alpha\alpha\alpha 20RC_{27}$, $\alpha\alpha\alpha 20RC_{28}$ and $\alpha\alpha\alpha 20RC_{29}$ are absolutely dominated by C_{28} and C_{29}, very low C_{27} sterane content (<10%) (Table 3-6 and Figure 3-8a). Biomarker features of type I crude oil have very good correlation with the source rock in Permian Fengcheng Fm, showing that this type of crude oil comes from the source rock in Permian Fengcheng Fm.

Type II crude oil has lower sterane/hopane ratio (0.28) and higher Ts/Tm ratio (on average 0.81), a small amount of rearranged sterane, $C_{29}>C_{28}>C_{27}$ in inverse "L" shape distribution, and a C_{27} content of 21% (Table 3-6 and Figure 3-8b), different from the features of Permian source rock to some extent, which indicate that this type of crude oil may be mixed with some crude oil generated by Jurassic dark mudstone.

2) Permian crude oil

The peak carbon numbers of the saturated hydrocarbons in Permian crude oil are C_{17} and C_{23}, and dominated by low carbon numbers less than C_{28}, which means that higher plants have less contribution to hydrocarbon generation materials. The crude oil has Pr/C_{17} and Ph/C_{18} of more than 1, and $Pr/Ph<1$, abundant β-carotane, showing the source rock deposited in a more reductive environment.

The Permian crude oil has higher tricyclic terpane content (with a tricyclic terpane/hopane ratio of 0.11–0.41), and more C_{23} than C_{21}, higher gammacerane content (with a gammacerane/C_{30} hopane >0.21), and Ts<Tm. Its regular terpane is characterized by very low C_{27} terpane content (<10%), high C_{28} terpane content, and near "Γ" shape distribution of C_{27}, C_{28} and C_{29} terpane (Table 3-6 and Figure 3-8c). It has little rearranged sterane and 4-methyl-sterane, and low maturity: with maturity of $\alpha\alpha\alpha C_{29}$ sterane 20S/(20S+20R) and C_{29} sterane $\beta\beta/(\alpha\alpha+\beta\beta)$ of 0.18–0.29 and 0.17–0.27, respectively. Oil-source correlation indicates that this kind of crude oil comes from the source rock in Permian Fengcheng Fm.

3) Carboniferous crude oil

In Carboniferous crude oil, normal paraffin hydrocarbons have been almost depleted by microbes (such as bacteria), and the isoprenoid alkanes are also significantly reduced, while the abundance of some compounds (such as sterane, terpane, gammacerane and β-carotane, etc.) that are more resistant to degradation become higher, and no 25-norhopane series has been tested out, so its degradation is at slight to moderate levels.

It has some tricyclic terpane (with tricyclic terpane/hopane ratio of 0.26–0.34), and more C_{23} than C_{21}, and higher gammacerane content (with a gammacerane/C_{30} hopane ratio of 0.34–0.83). In this kind of crude oil, the regular terpane $\alpha\alpha\alpha 20RC_{27}$, $\alpha\alpha\alpha 20RC_{28}$ and $\alpha\alpha\alpha 20RC_{29}$ have very apparent near "Γ" shape distribution, with very low C_{27} terpane content (<10%), the maturity value of $\alpha\alpha\alpha C_{29}$ sterane 20S/(20S+20R) and C_{29} sterane $\beta\beta/(\alpha\alpha+\beta\beta)$ are 0.27–0.39 and 0.32–0.36, respectively, at mature stage (Table 3-6 and Figure 3-8d). The biomarker features of this kind of crude oil are comparable with Permian source rock.

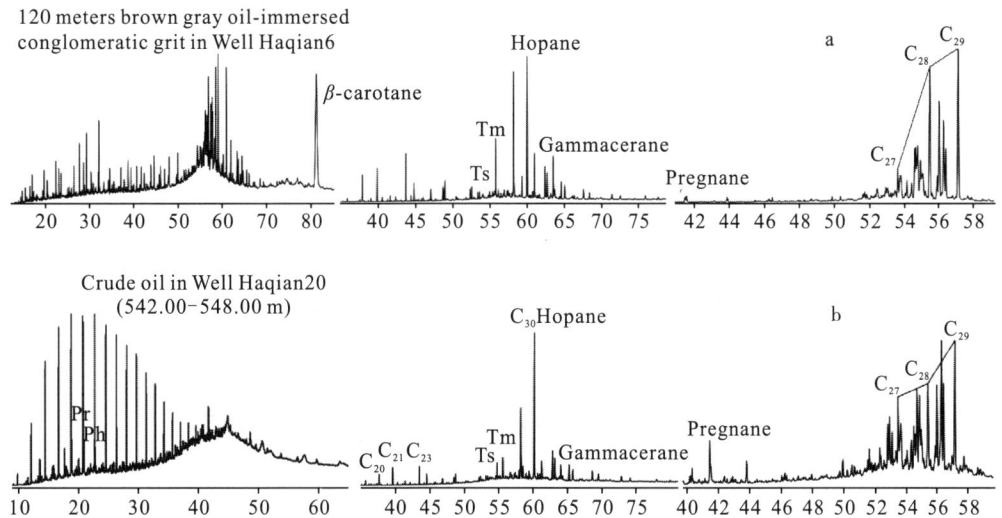

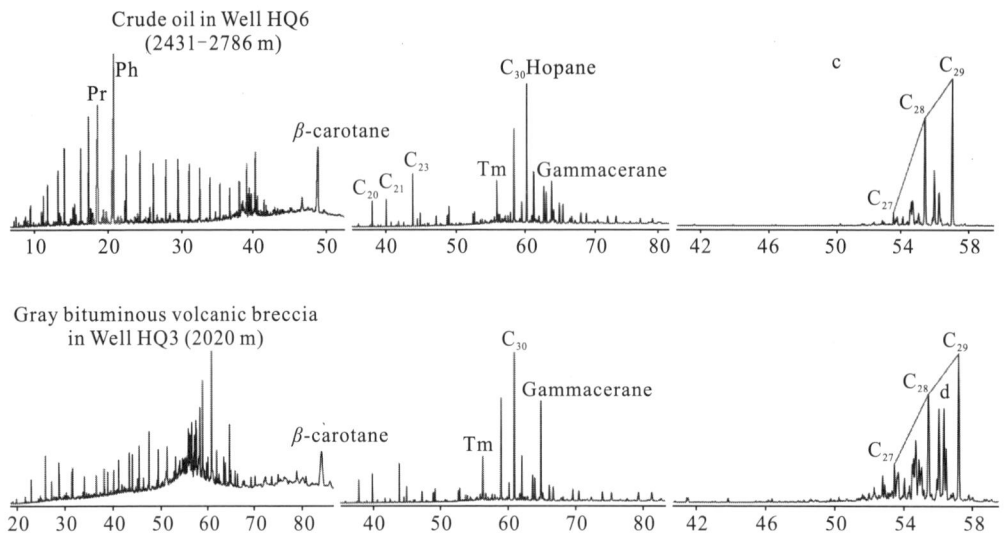

Figure 3-8 Characteristic spectrums of total ion current (TIC), terpane (m/z191) and sterane (m/z217) in crude oil/oil sand of the Hala'alate Mountain area.

Table 3-6 Statistics on biomarker parameters of crude oil in the Hala'alate Mountain area.

Well	Strata	$C_{29}S/(S+R)$	$C_{29}\beta\beta/\Sigma C_{29}$	C_{27}	C_{28}	C_{29}	Gammacerane $/C_{30}\alpha\beta$	Tricyclic /pentacyclic terpane	Sterane /hopane	Ts/Tm
HQ6	J oil sand	0.41	0.37	7	44	48	0.62	0.55	1.34	0.18
HQ6	J oil sand	0.38	0.33	6	46	48	0.42	0.33	0.90	0.10
HQ6	J oil sand	0.38	0.32	7	46	47	0.39	0.32	0.83	0.10
HS3	J asphalt	0.38	0.44	21	38	41	0.40	0.57	0.50	0.30
HQ20	J crude oil	0.38	0.51	22	35	43	0.12	0.20	0.28	0.81
HQ6	P	0.18	0.27	13	39	48	0.21	0.15	0.89	0.26
HQ6	P	0.24	0.20	3	46	51	0.37	0.11	6.99	0.21
HQ6	P	0.27	0.18	3	47	50	0.38	0.29	3.48	0.10
HQ6	P	0.27	0.17	3	48	50	0.48	0.41	4.50	0.07
HS1	P	0.29	0.18	3	46	51	0.63	0.34	4.18	0.13
HQ6	C	0.39	0.32	8	46	46	0.49	0.33	0.58	0.10
HQ6	C	0.39	0.33	7	46	47	0.35	0.32	1.39	0.12
HQ6	C	0.38	0.33	6	47	47	0.34	0.34	1.62	0.11
HQ102	C	0.27	0.23	5	43	52	0.83	0.32	1.07	0.09
HQ3	C	0.37	0.36	8	44	49	0.82	0.26	0.55	0.09

3. Analysis and correlation of oil-source

According to the biomarker features of source rocks and crude oil, the crude oil in the Hala'alate Mountain area has high comparability with the source rock in the Fengcheng Fm, thus it can be concluded that the crude oil in the Hala'alate Mountain area comes from the source rock in the Fengcheng Fm. However, the crude oil of different layers are quite different in maturity. The crude oil in Carboniferous system and the Fengcheng Fm in thrust nappe are apparently lower in maturity than that in Jurassic system and Carboniferous system in shallow onlap-denudation belt (Figure 3-9), which indicates that the crude oil in thrust nappe belt in the Hala'alate Mountain area and that in shallow onlap-denudation belt come from different sources. In order to further confirm oil source, the biomarkers in the source rocks in the Mahu Sag and the Hala'alate Mountain area have been analyzed and correlated in detail in this research.

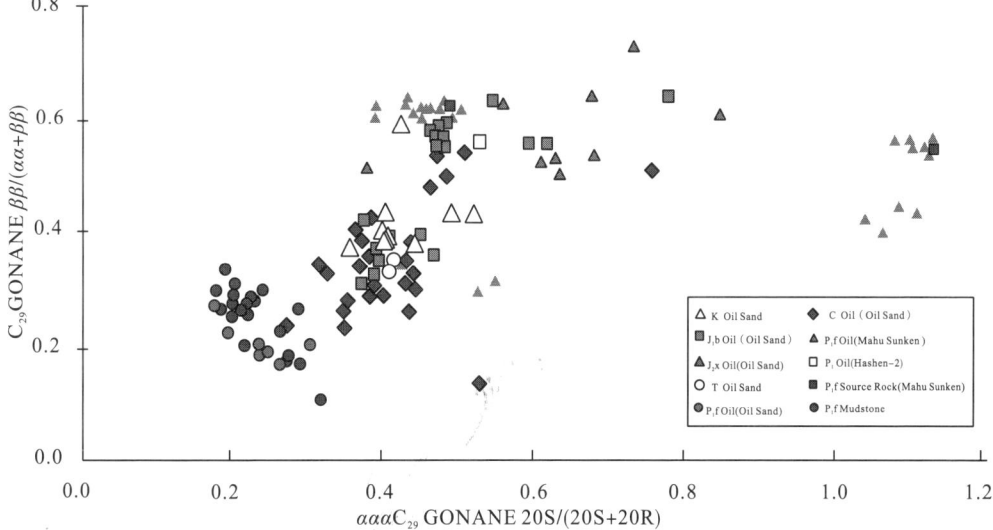

Figure 3-9 Scatter diagram of $\alpha\alpha\alpha C_{29}$ sterane 20S/(20S+20R) and C_{29} sterane $\beta\beta/(\beta\beta+\alpha\alpha)$ in crude oil in the Hala'alate Mountain area.

ETR index [ETR= (C_{28} tricyclic terpane + C_{29} tricyclic terpane)/(C_{28} tricyclic terpane + C_{29} tricyclic terpane + Ts)], firstly proposed by Holba, is used as chronologic parameter to distinguish crude oil from Triassic system and Jurassic system. This parameter has attracted the attention of researchers, once proposed (Tian et al., 2011; Ohm et al., 2008). By studying lacustrine source rocks in the Bozhong Sag in the BohaiBay Basin and the Junggar Basin, Hao Fang et al. found that ETR had good positive correlation with gammacerane/C_{30} hopane and homohopane index (HHI), etc., but negative correlation with Pr/Ph. It can be concluded that in lacustrine depositional environment, ETR index can be used to reflect depositional medium conditions (Hao et al., 2009). Meanwhile, HHI is usually used to reflect depositional environment: high HHI $C_{35}/\sum(C_{31}-C_{35})$ is regarded as a mark of marine depositional environment and inland saline lacustrine depositional environment (Chen et al., 2011).

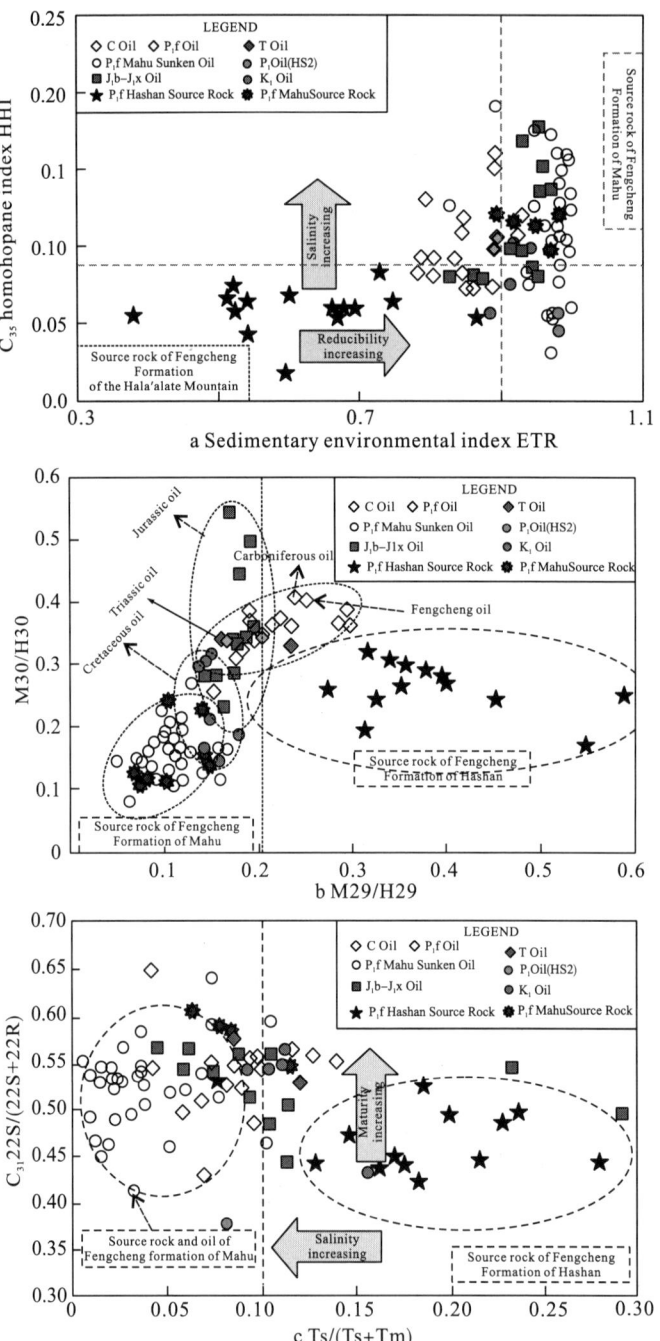

Figure 3-10 Correlation graphs of biomarker parameters in source rocks in Mahu Sag and the Hala'alate Mountain thrust nappe.

It can be seen from the correlation diagram between ETR and HHI that the source rock in the Fengcheng Fm in the Hala'alate Mountain area is characterized by low ETR (0.50-0.90) and low HHI (<0.08), while the source rock in the Mahu Sag has high ETR (0.90-0.98) and high HHI (>0.75) (Figure 3-10a and b), which indicates that the former had shallower water body and lower salinity when deposition than the later. Moreover, deep water body and high salinity environment can also suppress the conversion between some isomeric monomers (such as the conversion from Tm to Ts, and from moretane to hopane), thus the maturity parameters [Ts/(Ts+Tm), C_{29} moretane/C_{29} hopane, C_{30} moretane/C_{30} hopane] in the Mahu Sag are smaller than that in the Hala'alate Mountain area (Figure 3-10b and c).

According to the correlation between crude oil and source rocks, there are better correlation between Permian crude oil in the Mahu Sag, Jurassic and Cretaceous crude oil in shallow onlap-denudation belt in the Hala'alate Mountain area, and the source rock in the Fengcheng Fm in the Mahu Sag, so these crude oils mainly come from the source rock in the Fengcheng Fm in the Mahu Sag; while the crude oil in Permian and Carboniferous in the Hala'alate Mountain thrust nappe mainly come from the source rock in the Fengcheng Fm beneath the Hala'alate Mountain thrust nappe, and some crude oil samples mix with the crude oil generated by the source rock in the Mahu Sag.

3.3 Migration pathways and tracing of crude oil

Migration pattern of oil and gas in large oil and gas reservoirs has been studied by many researchers for many years. During the past few years, the development and perfection of nitrogenous compounds technology in crude oil allow us to get a deeper understanding on oil and gas migration. On the basis of analyzing oil and gas conducting systems and evolution, using the parameters that can indicate oil and gas migration to trace their migration pathways is of great significance for finding out oil and gas distribution pattern and selecting favourable exploration targets.

3.3.1 Conducting system of crude oil

1. West Hala'alate Mountain area

Before the Hala'alate Mountain thrusting in late Permian, the Upper Permian and Lower Permian Fengcheng Fm were shallower in burial depth (between 2000–2500 m). According to the relationship between physical properties and burial depth of reservoirs in the Wuxia area, the Fengcheng reservoirs during this period, with a porosity of 10%–15%, were fairly good in physical properties. The thrusting faults formed during this period acted as vertical conduits, hydrocarbons migrated along the faults vertically first and then migrated laterally along the sand bodies, mainly representing "T" shape conducting mode (Figure 3-11).

Thrust napping was violent in the Hala'alate Mountain West area at the end of Permian period. External napping system was the Carboniferous igneous rock and para autochthonous dolomitic rock; influenced by tectonic reformation, this area had many faults, and suffered long-term weathering and leaching reformation. The source rocks were situated beneath the thrust nappe, and hydrocarbons migrated along thrust faults and micro fractures (dissolved fractures) in "irregular treelike" pattern (Figure 3-12). The Triassic, Jurassic and Carboniferous systems overlap on the Hala'alate Mountain thrust nappe, like several sets of thick blanket sand layers, forming "fault-blanket conducting framework" with the oil-source faults in Wuxia fault belt in the south of the study area.

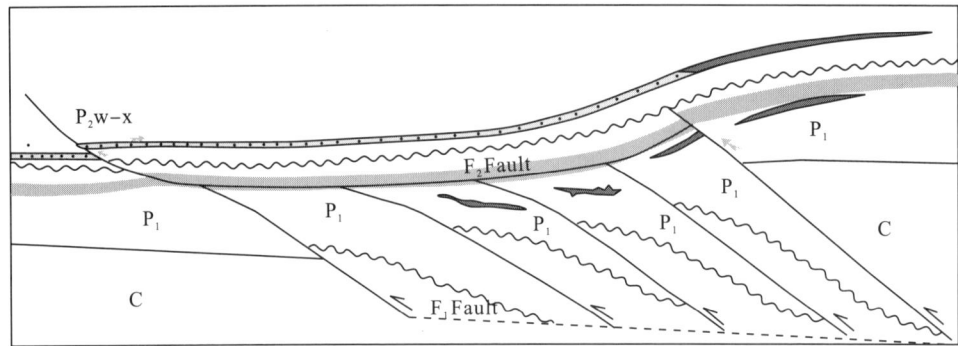

Figure 3-11　Hydrocarbon conducting mode in the Hala'alate Mountain West area during late Permian.

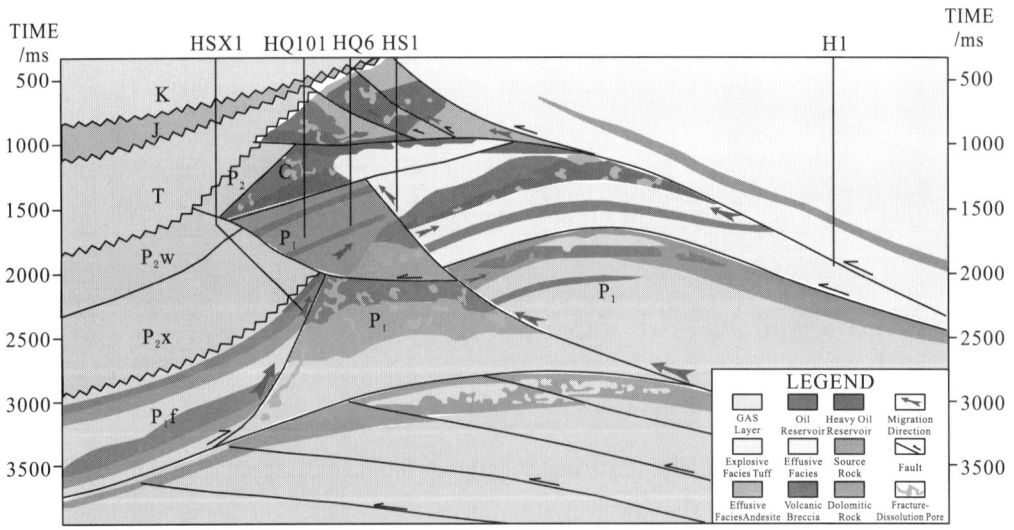

Figure 3-12　Hydrocarbon conducting mode in the Hala'alate Mountain West area.

2. East Hala'alate Mountain area

During the depositional stage of Permian-Triassic, thrust napping became apparently weaker in the East Hala'alate Mountain area, mainly manifesting as postspread distribution of multiphasethrust faults, and Permian system shows stepped distribution. During this period, the burial depth was generally 1500–2000 m; the sand bodies had better physical properties, thus faults and sand bodies could form step conduction mode (Figure 3-13).

In the end of Triassic period, tectonic compression had bigger influence on the East Hala'alate Mountain area, leading to strong deformation of the Triassic system and the formations below, and formation of a series of fault-related folds; and causing the sand bodies in Triassic and Upper Permian to distribute in concave-convex or residual style on profile, and drop of the lateral conducting capacity of these sand bodies to local lateral conduction. The shallower Jurassic and Cretaceous formations had weaker tectonic deformation, featuring overall tectonic subsidence, of slope tectonic setting, in which the sand bodies had good conducting capacity.

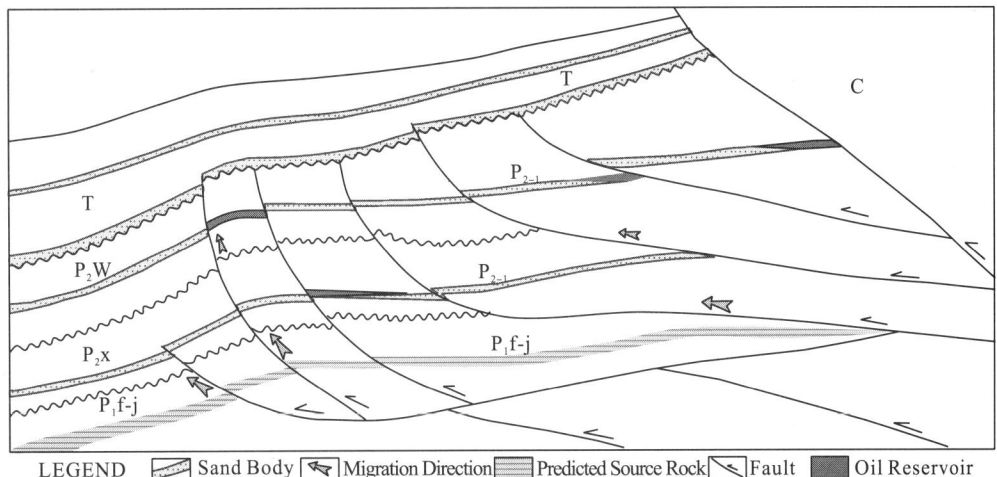

Figure 3-13 Step conducting mode in East Hala'alate Mountain area during Triassic period.

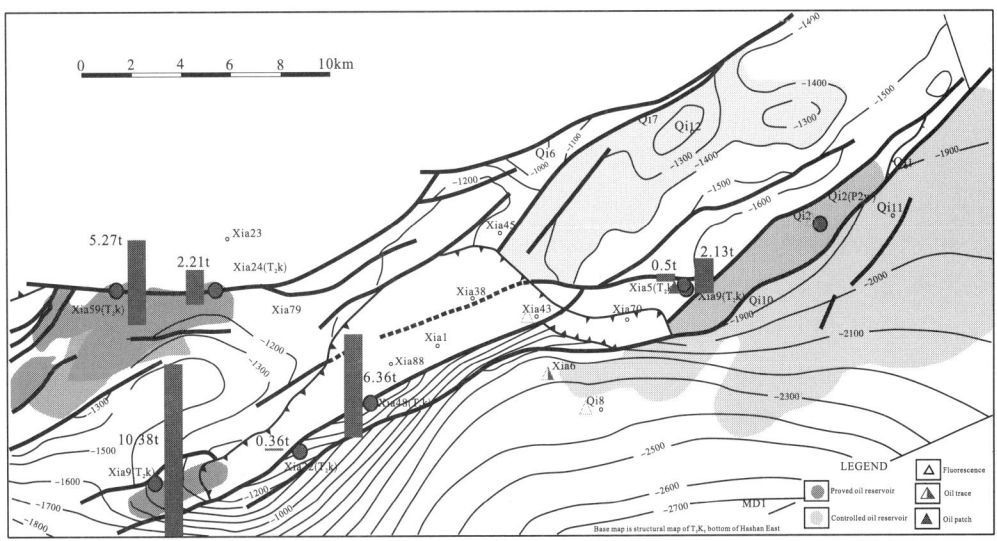

Figure 3-14 Distribution feature of oil and gas along faults in Hongqiba area.

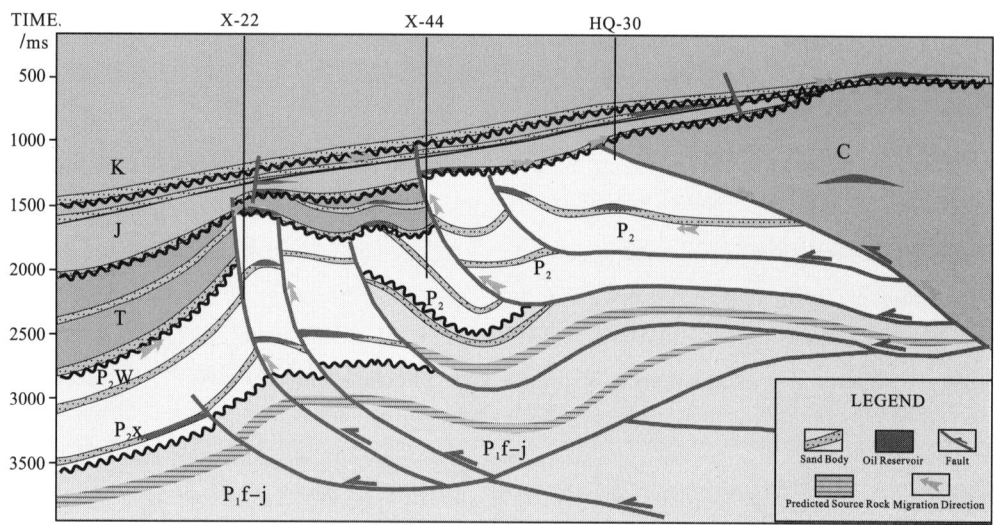

Figure 3-15 Hydrocarbon conducting mode in East Hala'alate Mountain area.

During later Jurassic and late Cretaceous periods, some faults (Xiahongnan, Wulanlinge, etc.) revived, and had better vertical conducting capacity. Furthermore, according to the analysis of hydrocarbon distribution features revealed by drilled wells, hydrocarbons in Well X9-X22-X48-Q5-Q2, and X59-X24 are chiefly distributed near faulted belts along Xiahonghan and Wulanlinge faults (Figure 3-14), which reflects lateral conduction of faults to some extent. In the East Hala'alate Mountain area, the deep and shallow layers have different conducting features: the shallow layers take fault-blanket conduction pattern, while the deep layers take vertical and lateral fault conduction pattern, with limited lateral conduction of sand bodies (Figure 3-15).

Generally, as the Hala'alate Mountain napping strength and way are different in different structural locations, different tectonic deformation response modes occur in different structural locations, which then controlled hydrocarbon conduction ways and conduction configuration styles in various periods and various locations. Before violent thrusting in late Permian, sand bodies in the Hala'alate Mountain tectonic belt had better lateral conducting capacity, and faults and sand bodies formed "T" shape conduction framework together. In the period before the deposition of Jurassic system and after basic finalization of the Hala'alate Mountain area, because of the differences in geologic structures, deep and shallow layers had two types of hydrocarbon conduction modes: in the deep layers, influenced by tectonic reformation, faults were well-developed, and the lateral conducting capacity of sand bodies was limited, so the main conducting mode is fault conduction; in shallow layers, stably distributed sand bodies and faults formed "fault-blanket conducting framework". The differences in conducting mode led to differences in hydrocarbon accumulation in different regions.

3.3.2 Tracing of crude oil migration pathways

In order to further figure out the dominant hydrocarbon migration paths in the Hala'alate Mountain area, from the regional perspective, we took the discovered oil reservoirs as study objects, used organic geochemical analysis as the tool, to trace and identify hydrocarbon migration paths, based on the geologic chromatographic effect (i.e., molecular structure difference) of nitrogenous compounds and biomarkers etc. in crude oil during migration.

Although accounting for a very small portion in crude oil, because of their special property (such as polarity), nitrogenous compounds can have direct and strong interactions with water, solid organic matter and minerals, so their distribution, composition characteristics and abundance changes can act as important evidences for tracing oil source (Ohm et al., 2008).

Currently, the study on nitrogenous compounds mainly focus on pyrrolic nitrogenous compounds, and the study results have achieved better application effect in hydrocarbon migration studies in continental petroliferous basins of China. Carbazole is a type of pyrrolic nitrogenous compounds, existing in non-hydrocarbons, and usually used to indicate hydrocarbon migration directions in special regions or oilfields. The carbazole molecules with nitrogen atom heterocycles have very strong polarity; the hydrogen atoms can bond with nitrogen atoms and negative atoms (such as oxygen atoms) in organic matter or clay minerals in formations to constitute hydrogen bonds, so some carbazole molecules would remain in conducting layers or reservoir beds, thus geologic chromatographic effect of the carbazoles would occur along hydrocarbon migration paths (Han et al., 2006; Zheng et al., 2004; Li et al., 1995; Dorbon, 1984), which manifests as: with the increase of migration distance, ①absolute concentration of the carbazoles decreases; ②alkyl carbazoles are richer than alkyl benzo carbazole; ③shielding carbazoles (1, 8-dimethyl carbazole, 1, 8-DMC) are richer than semi shielding carbazoles (C-1 and C-8 have only one alkyl substituent, such as 1- methyl carbazole, 1, 3-dimethyl carbazole, etc.) and bare carbazoles (such as 2, 7-dimethyl carbazole, 2, 7-DMC).

According to crude oil biomarker features, crude oil migration mode in different layers in the West Hala'alate Mountain area have apparent differences, migration distance of the crude oil in J_1b, J_2x and K are much larger than that of the crude oil in C and P (Figure 3-9). Based on geologic setting in this study area, before the deposition of Triassic and Jurassic systems, the study area experienced long-term erosion and deplanation, thus mudstone in unconformable weathering crust at Triassic bottom and Jurassic bottom are widespread, and thick lacustrine mudstone is developed in Upper Triassic Baijiantan Fm. Moreover, the unconformable basal conglomerate deposited quickly in the Badaowan Fm in proximal source style, with poor physical properties. All the three layers work together to separate the onlap-denudation belt and the thrust nappe belt. Meanwhile, according to the drilling results of Well HQ6 and HS1 etc., there is a widespread weathering hard crust 13–75 m below the unconformable surface at the Carboniferous top, the exotic napping system, with poorer reservoir properties, and almost no oil and gas shows, so there is little probability for crude oil in the onlap-denudation belt to flow back to the thrust nappe belt.

In addition, there are differences in parameters of nitrogenous compounds in crude oil in different units. The 1, 8-DMC/2, 7-DMC ratio and 1, 8-DMC/1, 5-DMC ratio of the crude oil in the onlap-denudation belt are higher than that in the thrust nappe belt, which also indicates that the crude oil in the onlap-denudation belt has undergone long distance migration (Figure 3-16), and the onlap-denudation belt and the thrust nappe belt belong to different hydrocarbon migration conducting systems.

Tracing of crude oil migration in the onlap-denudation belt: the onlap-denudation belt is generally a broad slope tectonic setting, where there develop thick and efficient blanket conducting sand bodies in wide and stable distribution, including Lower Cretaceous Qingshuihe Fm (K_1q), Middle Jurassic Xishanyao Fm (J_2x) and Lower Jurassic Badaowan Fm (J_1b); its downdip position matches with the oil source faults in southern Wuxia fault belt, forming favourable "fault-blanket" conducting framework for hydrocarbon migration, thus the hydrocarbons generated by the source rocks in the Mahu Sag can migrate along oil source faults-blanket sands-secondary faults-blanket sands in a "relay style", creating three oil-rich layers (K_1q, J_2x and J_1b) in the onlap-denudation belt (Figure 3-17).

In order to further study the hydrocarbon migration paths of crude oil in the efficient conducting sand bodies in the onlap-denudation belt, we utilized biomarker parameters and nitrogenous compound parameters in crude oil to trace crude oil migration features in Jurassic Badaowan Fm and Xishanyao Fm in the Hala'alate Mountain area. It can be seen that crude oil is mainly from Mahu Sag to the south, and the crude oil migrates dominantly along structural ridges (sand ridges) on plane (Figure 3-18).

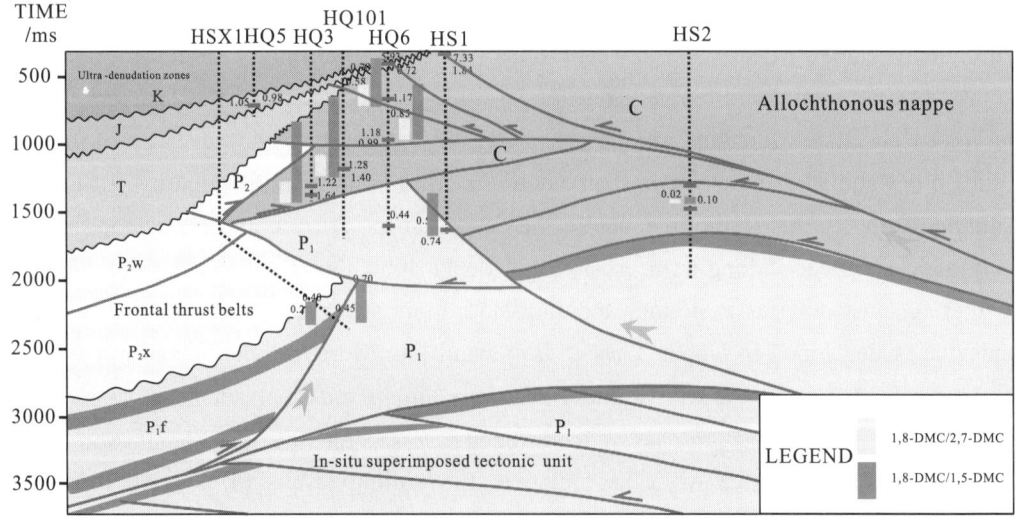

Figure 3-16　Profile of nitrogenous compound parameter distribution in various units of the Hala'alate Mountain area.

Chapter 3 Oil-source Correlation and Migration Pathway Tracing

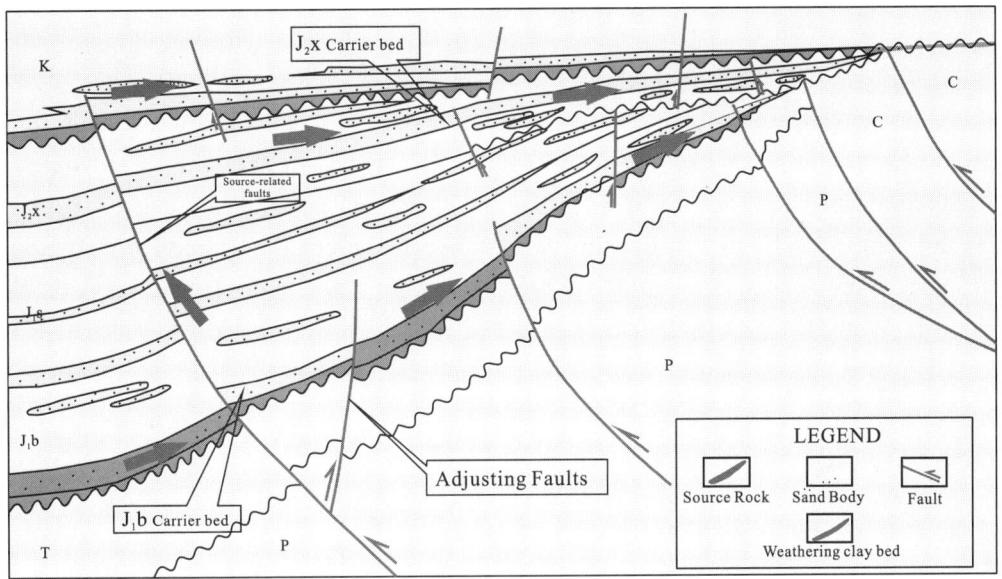

Figure 3-17 "Fault-blanket" conducting mode in onlap-denudation belt of the Hala'alate Mountain area.

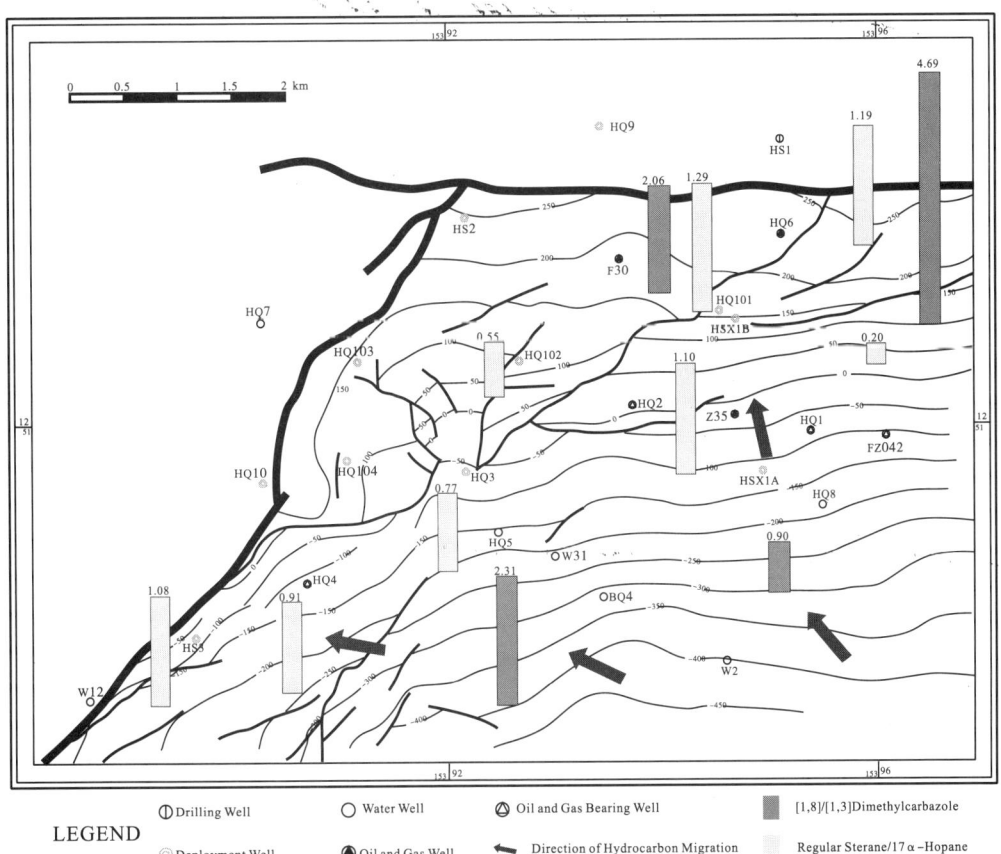

Figure 3-18 Hydrocarbon migration variation trend in West the Hala'alate Mountain area.

Tracing of crude oil migration in the thrust nappe belt: the thrust nappe belt shows multiphase thrust napping superimposition and multiphase imbricated thrusts, with complex geologic structure. The thrust nappe belt has many reverse slip faults and thrust faults which might provide pathways for hydrocarbon migration. The analysis results of types, fluorescence and homogenization temperature of inclusions in F3 faulted belt indicate that F3 fault was active for long period, and the fracture veins captured multiphase hydrocarbon inclusions. Systematic geochemical analysis of the oil-bearing sections (26 samples between 60.1–2786 m section) in Well HQ6 in the thrust nappe belt also shows several geochemical parameters reflecting hydrocarbon migration have regular variations from deep layers to shallow layers: regular sterane/17-α hopane ratio changes from 3.39→2.64→1.58→1.41→0.62, $\alpha\alpha\alpha C_{29}20S/(20S+20R)$ ratio changes from 0.20→0.25→0.37→0.39→0.41, tricyclic terpane/17-α hopane ratio changes from 0.05→0.10→0.15→0.31→0.33, (pregnane+homopregnane)/C_{29} sterane ratio changes from 0.004→0.011→0.018→0.020→0.043, all indicating that faults are important pathways for vertical hydrocarbon migration (Figure 3-19).

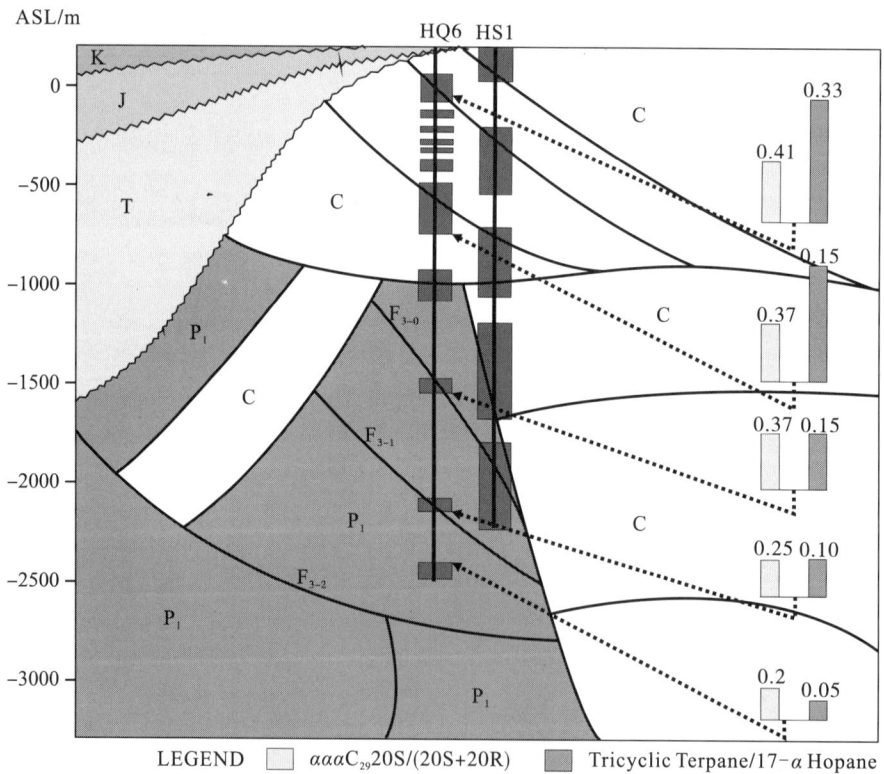

Figure 3-19 Vertical parameter variation profile of hydrocarbons in F3 faulted belt in West the Hala' alate Mountain area.

Chapter 4
Development Features of Reservoir Fractures

Fractures are important migration pathways and reservoir space for volcanic fractured reservoirs. They can be classified into structural fractures and nonstructural fractures according to their origin, in which, structural fractures are the direct product of tectonic stress. The tectonic movements of different geological periods in the same area have different tectonic stress nature, which leads to the difference in occurrence of faults and structural fractures formed in different tectonic periods. In the Hala'alate Mountain area located in northwestern mountain front thrust belt in the Junggar Basin, faults and structural fractures are extensively developed in the Carboniferous-Permian, due to multi-phase tectonic compressive nappe movements. It is the superposition of multi-phase tectonic movements of different stress properties that resulted in the complexity of structural fracture occurrence in the Hala'alate Mountain area. As it is difficult to get a comprehensive and thorough understanding on the overall development features of structural fractures in an area by using only one research means, several technological means have been taken to find out the occurrence, development period of structural fractures and their relationship with hydrocarbon accumulation in the area in this research, in the hope to provide reference and guidance for oil and gas exploration.

4.1 Development features of core fractures

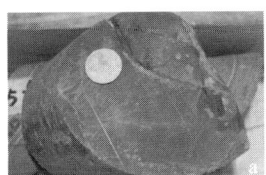

Straight high-angle fracture, mudstone (Well HS1, 2099.70 m).

High-angle fracture in the late stage with straight surface, mudstone (Well HQ101, 2237.75 m).

Figure 4-1 Typical features of core fractures in the Hala'alate Mountain area.

Based on observations of core taken from 8 wells (Well HS1, HS2, HQ3, HQ4, HQ6, HQ7,

HQ101 and HQ102) in the Hala'alate Mountain area, it is found that the unfilled fractures in the Carboniferous-Permian are large in dip angle (Figure 4-1), and there are also netted fractures of high filling degree in Permian, which is the direct evidence that the Hala'alate Mountain area underwent multi-phase strong tectonic movements.

4.1.1 Identification and statistic of fractures in coring sections

Structural fractures are formed under tectonic stress, but structural fractures associated with different phases of tectonic movements are different in occurrence and shape. As the Hala'alate Mountain area underwent multi-phase strong tectonic movements during geological history, seismic sections of this area show there develop multiple reverse faults in Carboniferous and Permian, and because of different faulting period and intensity, the structural fractures of multi-phase are obviously mutual intersecting (Figure 4-2), and different to some extent in dip angle. Generally it is considered that fractures formed in the same phase of tectonic movement have similar occurrence features, therefore, the formation and development phase of fractures can be figured out by studying different dip angles of structural fractures.

Observation of fractures in cores taken from wells in the research area shows high-angle fractures and vertical fractures account for a large proportion, low-angle fractures and horizontal fractures account for a small proportion, and there are also some netted fractures; high-angle fractures and vertical fractures can be seen in cores of all wells, especially in Well HS1, HQ6 and HQ101.

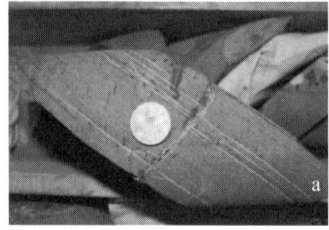

Development of two phases of structural fractures (Well HS1, 2096.40–2097.40m).

High-angle fracture in late stage cutting netted fracture in early stage (Well HQ101, 2237.59m).

Figure 4-2 Cutting relationship of structural fractures.

The statistic results of core fractures (Figure 4-3) show that fractures in Well HS1, HQ6 and HQ101 are dominated by high-angle oblique ones, and fractures in Well HS2 and HQ6 are dominated by low-angle ones; on the whole, structural fractures in the Hala'alate Mountain area are developed in all kinds of lithologies of Carboniferous and Permian, and primarily high-angle fractures and vertical fractures, secondarily low-angle fractures and horizontal fractures, with some netted fractures in local well intervals rich in high-angle fractures and vertical fractures.

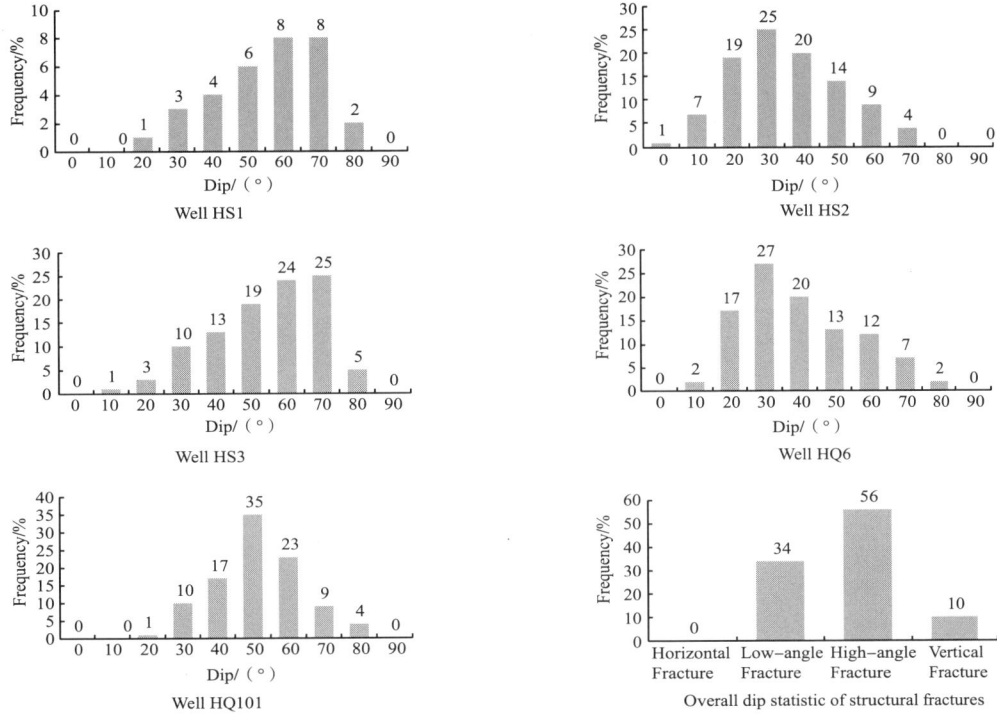

Figure 4-3   Histogram of distribution frequency of structural fracture dip angle.

4.1.2   Filling features of fractures in coring sections

Normally fractures in formations are all filled to different degrees. Therefore, observation and description of them are important for understanding the genesis, forming phase of them and their effectiveness for hydrocarbon migration and accumulation. Observation and description of cores taken from 8 wells in the Hala'alate Mountain area show the structural fractures are all filled to different degrees (Table 4-1), overall the fractures are average in effectiveness, some of the structural fractures are filled with calcite, the unfilled high-angle fractures and vertical fractures are filled with crude oil or asphalt, and dissolution of calcite can be seen, with the pores after dissolution usually filled with asphalt.

Observation and comparison of a large number of cores show that the structural fractures of different occurrence differ widely in filling degree. The netted structural fractures, low-angle structural fractures and several high-angle structural fractures of small width are almost fully filled with calcite vein (Figure 4-4a, b, and c); while high-angle structural fractures with larger width and vertical structural fractures are filled at a low degree, mainly unfilled or half-filled by calcite vein, and local dissolution can be seen in the calcite vein (Figure 4-4d). The core sections with high-angle structural fractures or vertical fractures filled at low degree usually have fairly good oil and gas shows (Figure 4-4e and f), indicating this kind of structural fracture is higher in effectiveness.

Table 4-1　Filling conditions of fractures in Carboniferous strata.

Well No.	Filling conditions of fractures
HS1	The absolute majority are filled with calcite, some contain oil and asphalt
HS2	Calcite, quartz
HQ101	Filled with calcite, containing asphalt and oil
HQ102	Filled with calcite, containing asphalt and oil
HQ3	Filled with calcite and quartz, some filled with oil and asphalt
HQ4	Filled with calcite, some filled with quartz
HQ6	The absolute majority filled with calcite, some containing oil and asphalt
HQ7	Filled with calcite

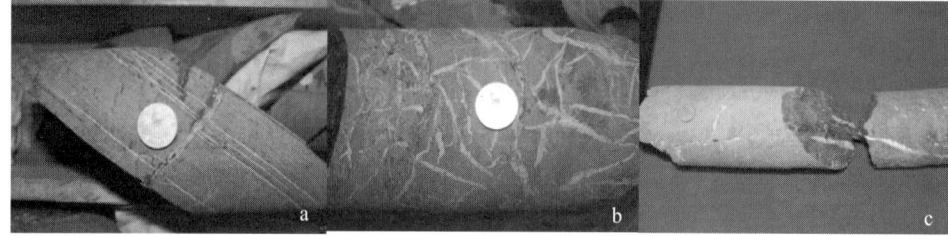

Calcite vein filling fractures dissolved in late stage(Well HS1, 2096.40–2097.40 m).　Netted fractures fully filled with calcite vein(Well HS1, 2153.7 m).　Vertical fractures half-filled with calcite vein(Well HQ6, 255.50 m).

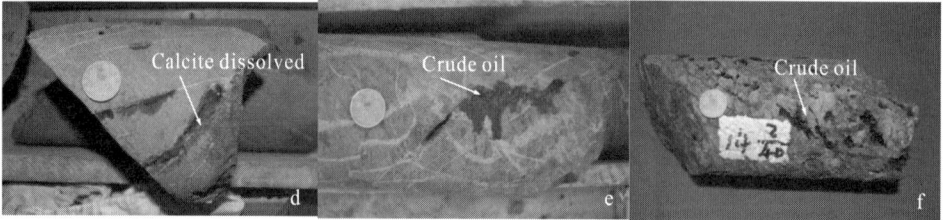

The place in a high-angle fracture where calcite is dissolved is filled with crude oil (Well HS1, 2099 m).　High-angle fractures filled with crude oil(Well HS1, 2152.50–2154.21 m).　High-angle fractures filled with crudeoil(Well HQ101, 1737.70 m).

Figure 4-4　Filling features of fractures in cores of Carboniferous-Permian.

4.2　Imaging log responses of fractures

Imaging logging is a method to conduct geophysical parameter imaging of borehole wall and objects around well according to the observation of geophysical field in the borehole. Broadly speaking, imaging logging includes borehole wall imaging, near-well imaging and inter-well imaging. Images obtained from imaging logging can visually reflect development conditions of subsurface fractures and dissolved pores, therefore, the advent of imaging logging technology brings huge convenience to evaluation of special oil and gas reservoir space (Lu et al., 2004). As an advanced logging technology using the latest electronic technique and computer technology, imaging logging has the advantage of high resolution of data. Currently, the mainstream imaging logging tools in the world include Schlumberger MAXIS500, Halliburton EXCELL-2000 and Atlas ECLIPS-5700, etc. The last one was mainly used in the imaging logging operation in Carboniferous-Permian reservoir of the Hala'alate Mountain area, obtaining high-resolution XRMI images from Well HQ3, HQ6, HQ101, HS1 and HS2.

According to the genetic mechanism and opening degree, fractures interpreted from imaging logging usually can be classified into three types, high-conductivity fractures, high-resistivity fractures and induced fractures. The imaging logging data and core description correlation of five typical wells (Well HS1, HS2, HQ3, HQ6 and HQ101) in the Hala'alate Mountain area indicate that different types of fractures show different features in imaging logging. Located in mountain front thrust belt of the Wuxia faulting zone, the Hala'alate Mountain area is widely developed with structural fractures in Carboniferous volcanic rock and Permian clastic rock. They are mainly high-angle fractures, accompanied by vertical fractures, and some low-angle fractures and netted fractures. The fracture occurrence recorded by imaging logging is basically consistent with that found by core observation. It is necessary to explain that the color changes of imaging logging pictures reflect changesin stratigraphic lithology and physical property around borehole wall, and cannot wholly correspond to the lithology of actual strata. This paper mainly discusses natural structural fractures, and the interference of fractures induced by strata boundary and drilling must be excluded, so they were not included in the statistics of imaging logging fractures.

(1) The high-conductivity fracture shows black high-conductivity anomaly in imaging logging images, and is characterized by dark strip similar to sine curve (Figure 4-5). When high-conductivity fractures are developed, they are shown as multi-groups of dark strips similar to sine. Fractures can also cut any medium and form intersecting fracture combinations.

(2) When filled with minerals like calcite and quartz, the high-resistivity fractures generally show bright high-resistivity anomaly (Figure 4-6). When a formation is faulted and the faulting zone is filled with high-resistivity minerals, the faulting zone is shown as high-resistivity bright spot.

(3) The induced fractures are induced during drilling, and in feather shape. The most distinct feature of induced fractures is that they appear in symmetry along borehole wall (Figure 4-7), so the trend of induced fractures can better reflect the direction of current maximum horizontal stress.

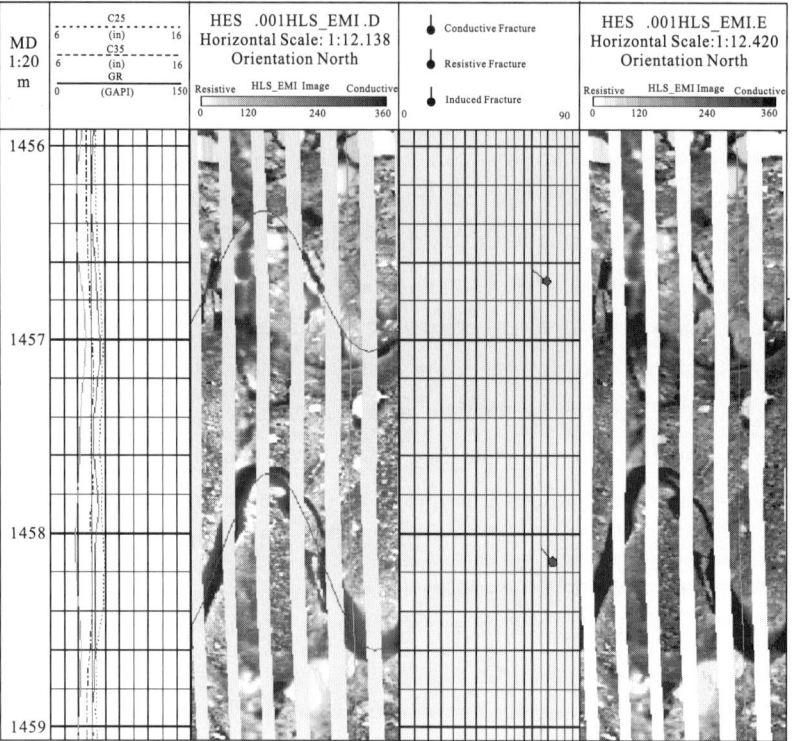

Figure 4-5 Imaging features of high-conductivity fractures in XRMI image, well HQ6 (1456–1459 m).

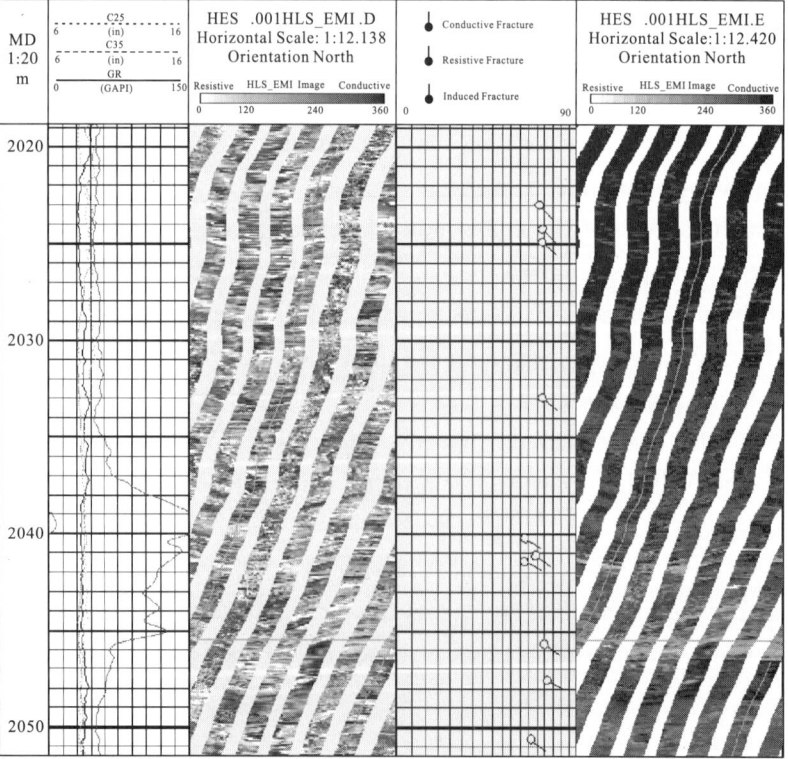

Figure 4-6 Imaging features of high-resistivity fractures in XRMI image, Well HQ6 (2020–2050 m).

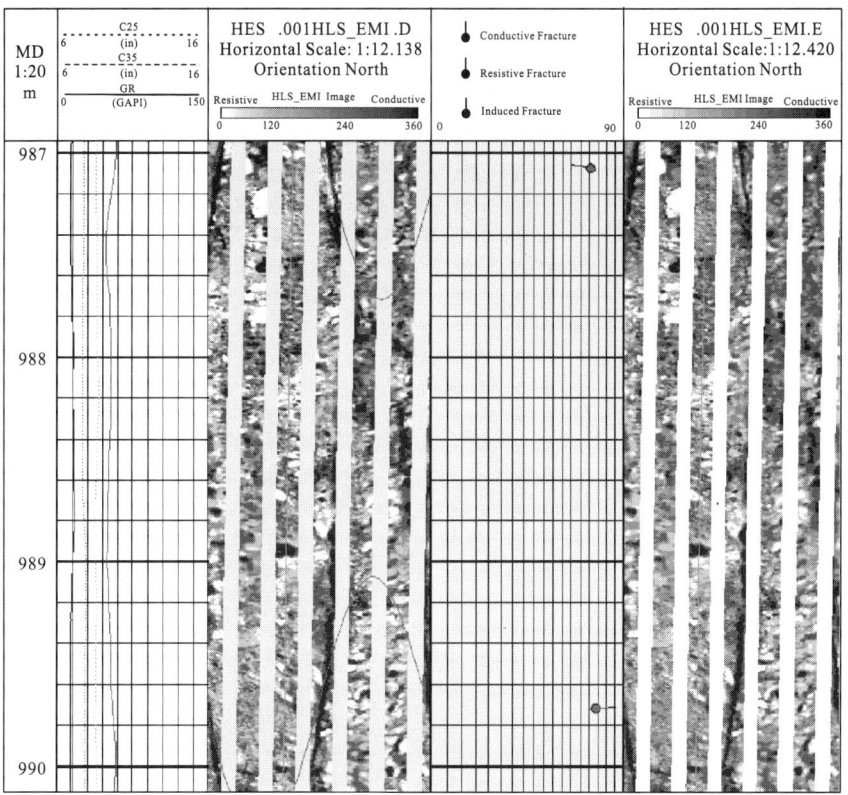

Figure 4-7　Imaging features of induced fractures in XRMI image, Well HQ6 (987–990 m).

4.2.1　Imaging logging response of typical drilling fractures

1. Well HS1

1) Development features of fractures

Fractures of Well HS1 in XRMI images include high-conductivity fractures, induced fractures and high-resistivity fractures.

The high-conductivity fractures are shown as low-resistivity dark sine strips in electronic imaging pictures, and there is usually dissolution along fractures, which is indicated by red tadpole on result map (Figure 4-8). From the statistics of fracture occurrence, we can see that the high conductivity fractures of this well are mainly middle-high angle fractures, with a dip angle of 12°–80°, in S-E trend and NE-SW strike.

Generated during the drilling process, drilling induced fractures, in feather shape, feature symmetric occurrence along borehole wall (Figure 4-9). The trend of induced fractures can better reflect the direction of current maximum horizontal stress. The induced fractures are 50°–88° in dip angle, N-W trend and NE-SW strike.

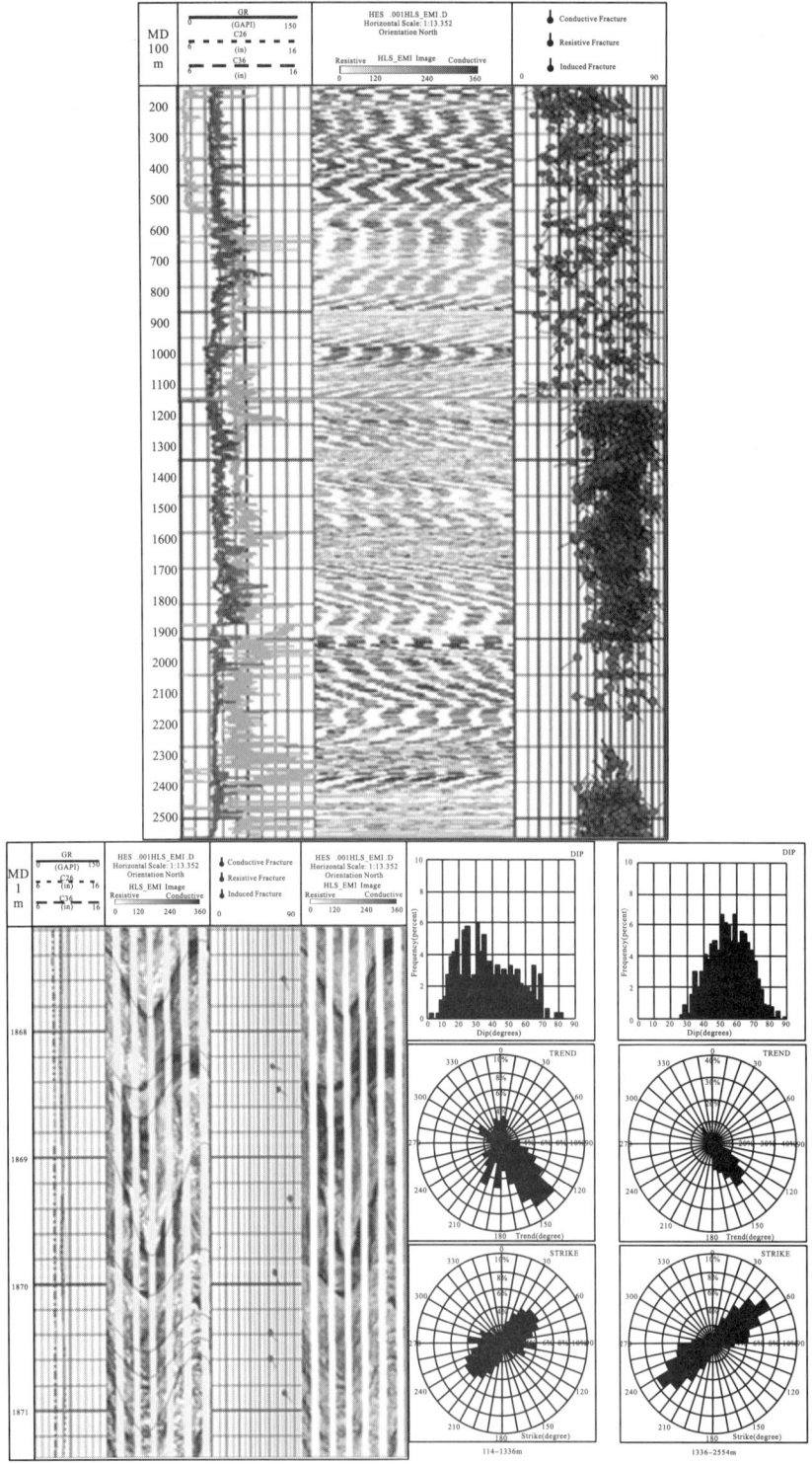

Figure 4-8 Features, distribution and occurrence of high-conductivity fractures in XRMI image, Well HS1 (114–2554 m).

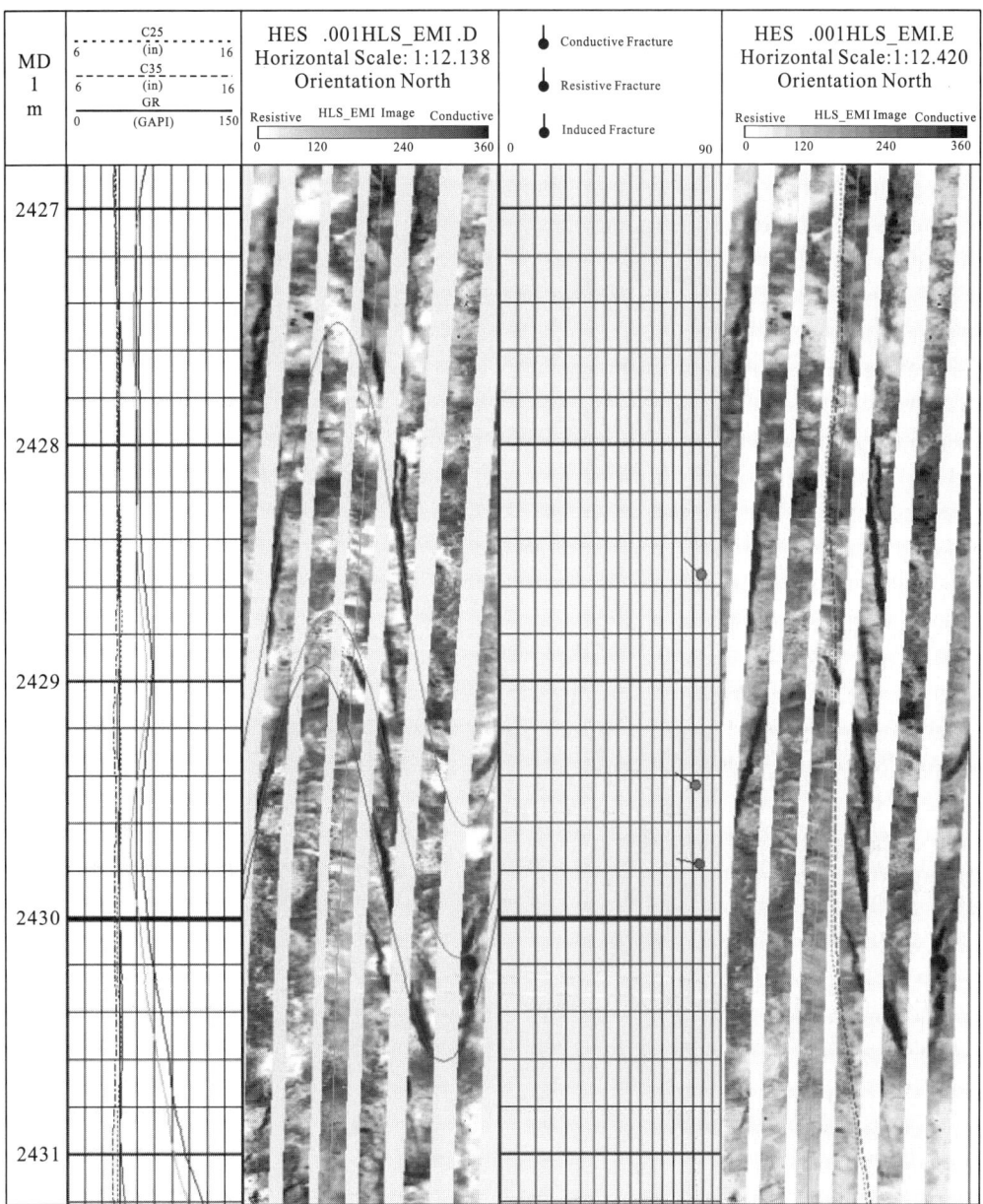

Figure 4-9 Imaging features and occurrence statistics of induced fractures in Well HS1 in XRMI image (2427-2431m).

The high-resistivity fractures are closed or filled, shown as bright yellow or white sine lines on result map. Usually, they are early fractures filled by cement or closed in the late stage (Figure 4-10). The occurrence statistics show that the high-resistivity fractures are 14°-70° in dip angle, and random in trend and strike.

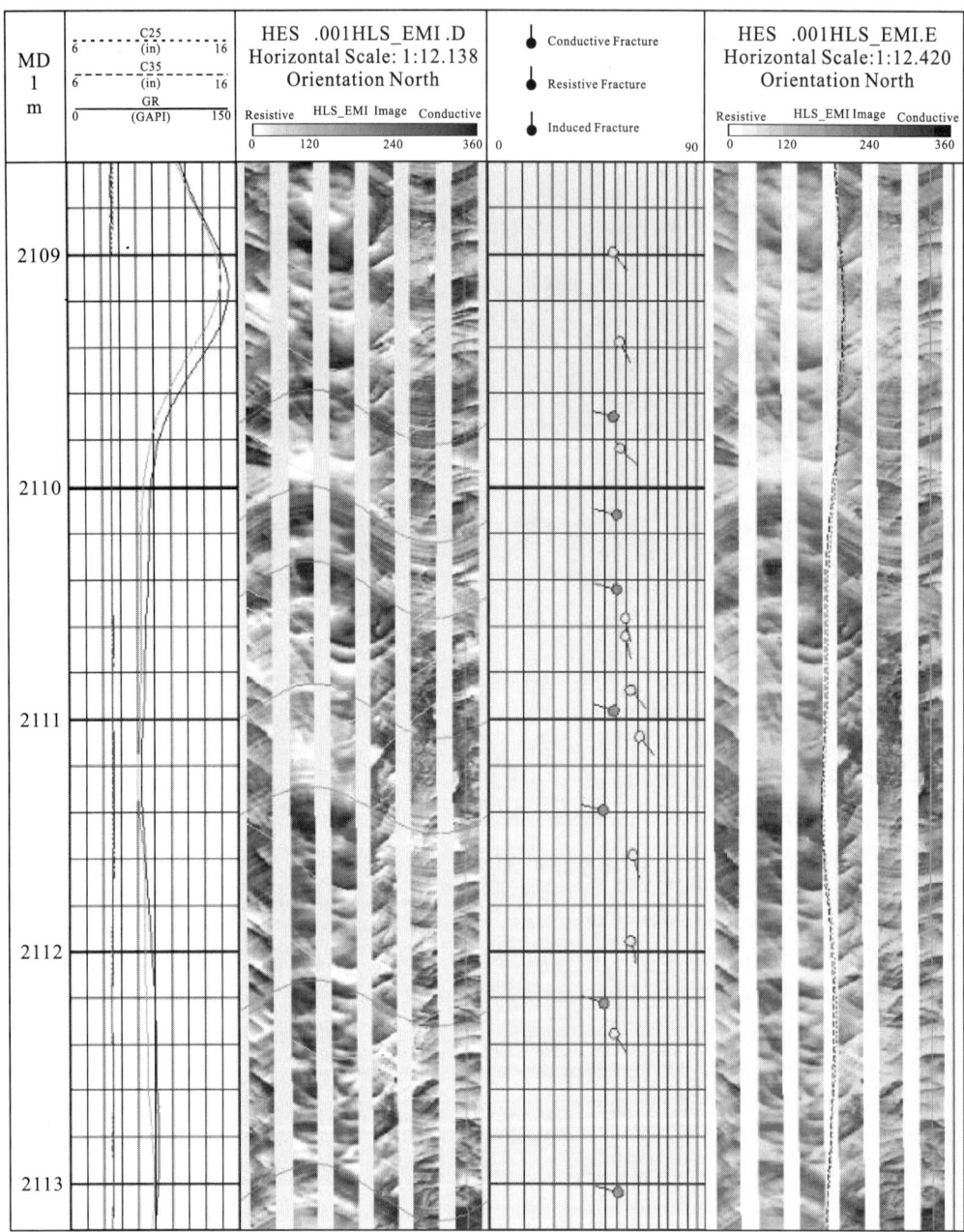

Figure 4-10 Distribution and occurrence statistics of high-resistivity fractures in XRMI, Well HS1 (2109–2113 m).

2) Quantitative calculation of high-conductivity facture parameters

The calculation of high-conductivity fractures parameters is to perform continuous statistics of fracture apparent parameters within the statistic window, and quantitative calculation is based on empirical formula from experiment and mathematical simulation:

$$W = a \cdot A \cdot R_{XO}^{b} \cdot R_{M}^{(1-b)}$$

Where, W is fracture width; A is the area of conductivity anomaly resulted by fractures; R_{XO} is formation conductivity (normally resistivity of invaded zone); R_M is mud resistivity; a and b are constants related to the instrument, and b is close to zero; A and R_{XO} are calculated based on the image calibrated to LLS.

The fracture porosity is expressed as:
$$VPA = \sum W_i \cdot L_i / (L \cdot \pi \cdot D)$$
Where, VPA is fracture porosity; W_i is the average width of the i fracture; L_i is the width of the i fracture in statistic window L (normally L is 1 m or 0.6096 m); D is caliper.

The following apparent fracture parameters of high-conductivity fractures can be obtained by quantitative calculation:

(1) Fracture density ($FVDC$) is the total number of fractures seen in every meter of well interval.

(2) Fracture length ($FVTL$) is the sum length of fractures seen in every square meter of borehole wall.

(3) Fracture hydrodynamic width (FVA) is the cubic root of sum of cube of fracture trajectory width.

(4) Fracture porosity ($FVPA$) is the ratio of borehole wall area occupied by fractures to that covered by imaging logging in every meter of well interval.

The quantitative calculation result of fracture parameters of Well HS1 indicates that ① well interval 114–1336 m has a fracture length <3.2 m/m², on average 1.9 m/m²; fracture hydrodynamic width <140 μm, on average 96 μm; fracture apparent porosity <0.005%, on average 0.00195%; ② well interval 1336–2554 m has a fracture length <6 m/m², on average 2.6 m/m²; fracture hydrodynamic width <120 μm, on average 63.8 μm; and fracture apparent porosity < 0.005%, on average 0.0019%. The reservoir of this well is mainly dissolution-pore reservoir.

2. Well HS2

1) Development features of fractures

Fractures of Well HS2 in images are mainly high-conductivity fractures, high-resistivity fractures and induced fractures.

The high-conductivity fractures in electronic imaging pictures are shown as low-resistivity dark sine strip curves, and dissolution is often seen along fractures, which is indicated by red tadpole on result map (Figure 4-11).

The high-resistivity fractures, closed or filled, are shown as bright yellow or white sine lines on result map (Figure 4-12), and usually early stage fractures filled with cement or closed in the late stage.

The statistics show that the high-conductivity fractures are mainly middle-high in angle, with larger openings and dip angle of 20°–80°, mostly in SSE dip and NEE-SWW strike

(Figure 4-13). The high-resistivity fractures have a dip angle of 10°–80°, trend mainly of SSE, NW and NNE, strike mainly of NE-SW, NWW-SEE and NEE-SWW (Figure 4-14).

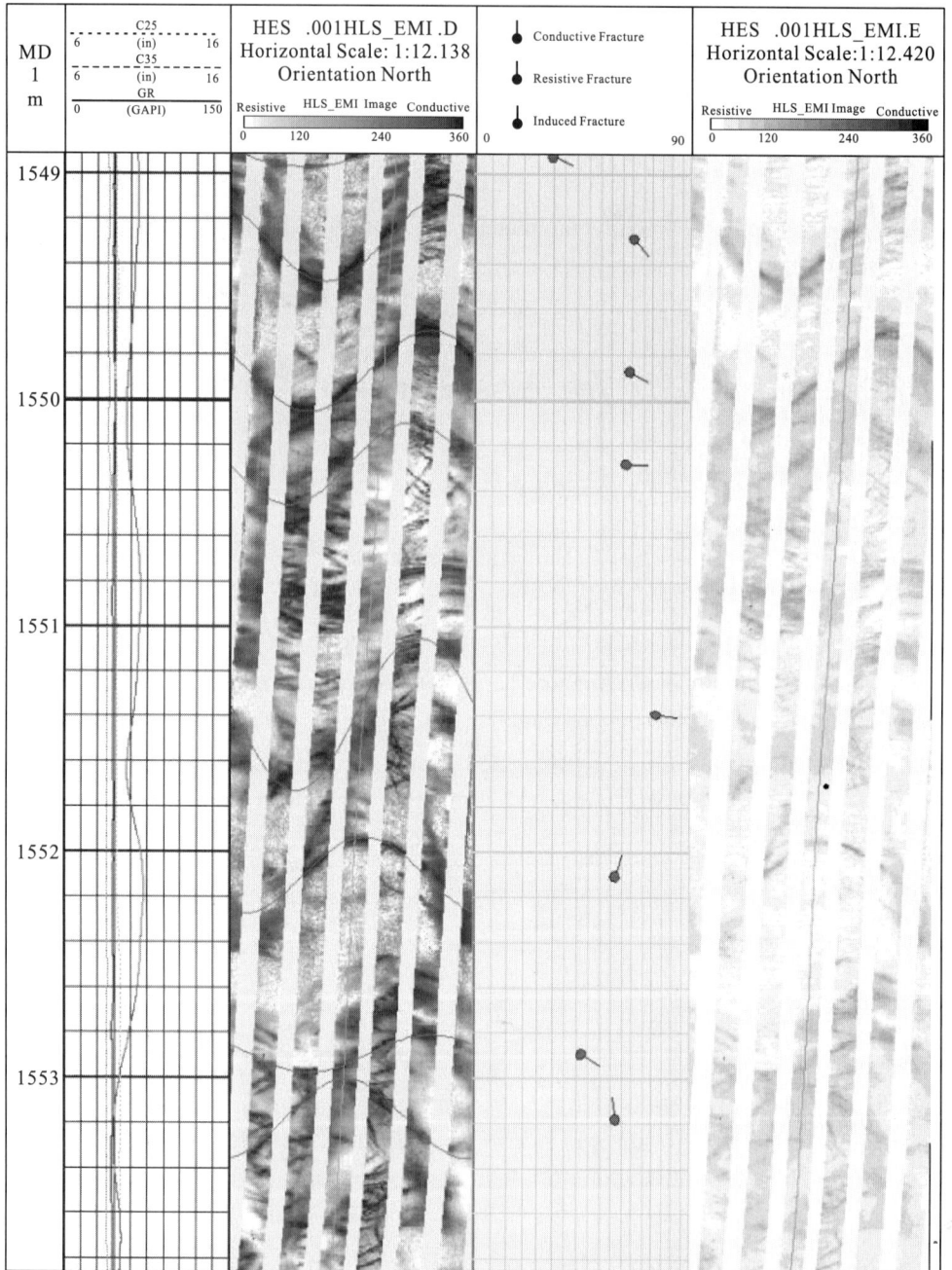

Figure 4-11 Imaging logging features of high-conductivity fractures in Well HS2 (1549–1553 m).

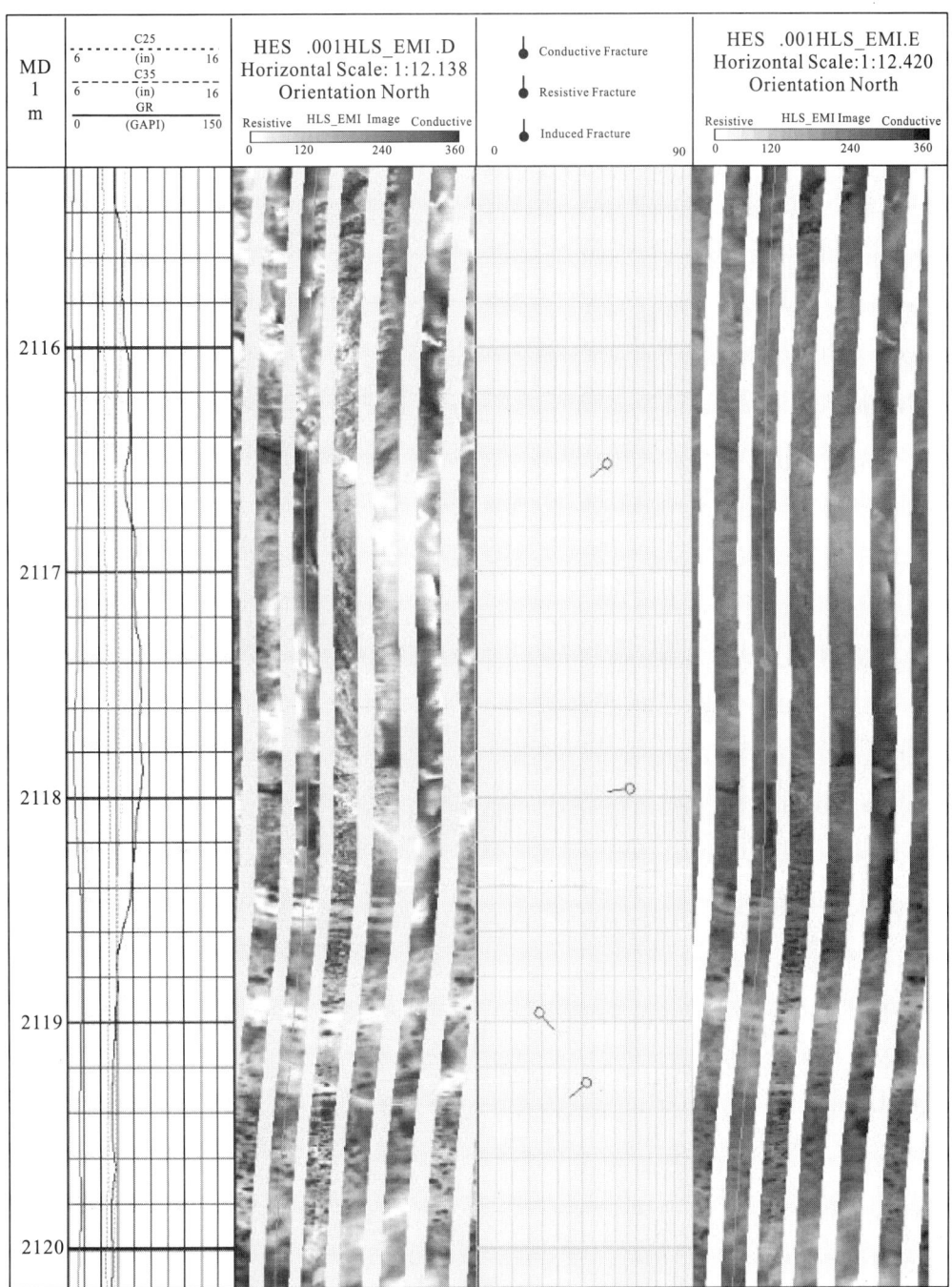

Figure 4-12　Imaging logging features of high-resistivity fractures in Well HS2 (2116–2120 m).

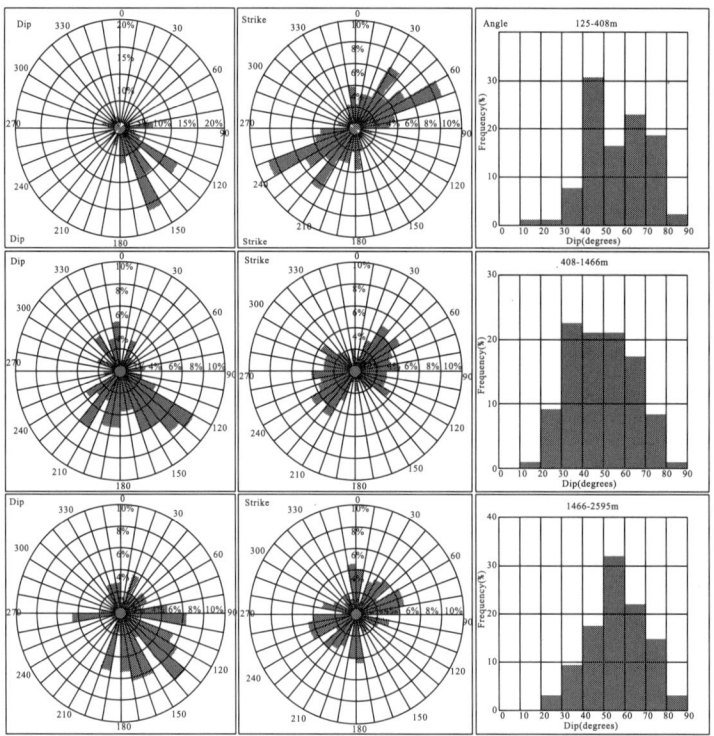

Figure 4-13　Occurrence statistics of high-conductivity fractures in Well HS2.

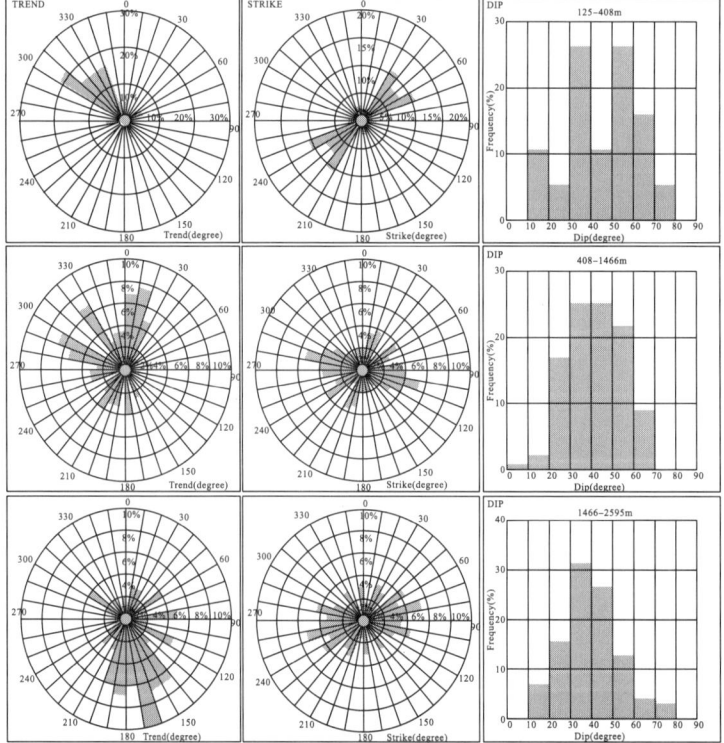

Figure 4-14　Occurrence statistics of high-resistivity fractures in Well HS2.

2) Quantitative calculation of high-conductivity facture parameters

The quantitative calculation result of fracture parameters (Figure 4-15, Figure 4-16) shows that fractures in Well HS2 have a length <5.722 m/m^2, on average 1.652 m/m^2; hydrodynamic width <0.304 mm, on average 0.120 mm; and apparent porosity <1.304%, on average 0.165%. Conventional analysis shows that most well intervals with better physical properties have different degrees of cavity enlargement, so imaging effect is poor and geological features like fractures are not clear at these intervals, thus it is inferred that most of the reservoirs in this well are dissolution-pore type.

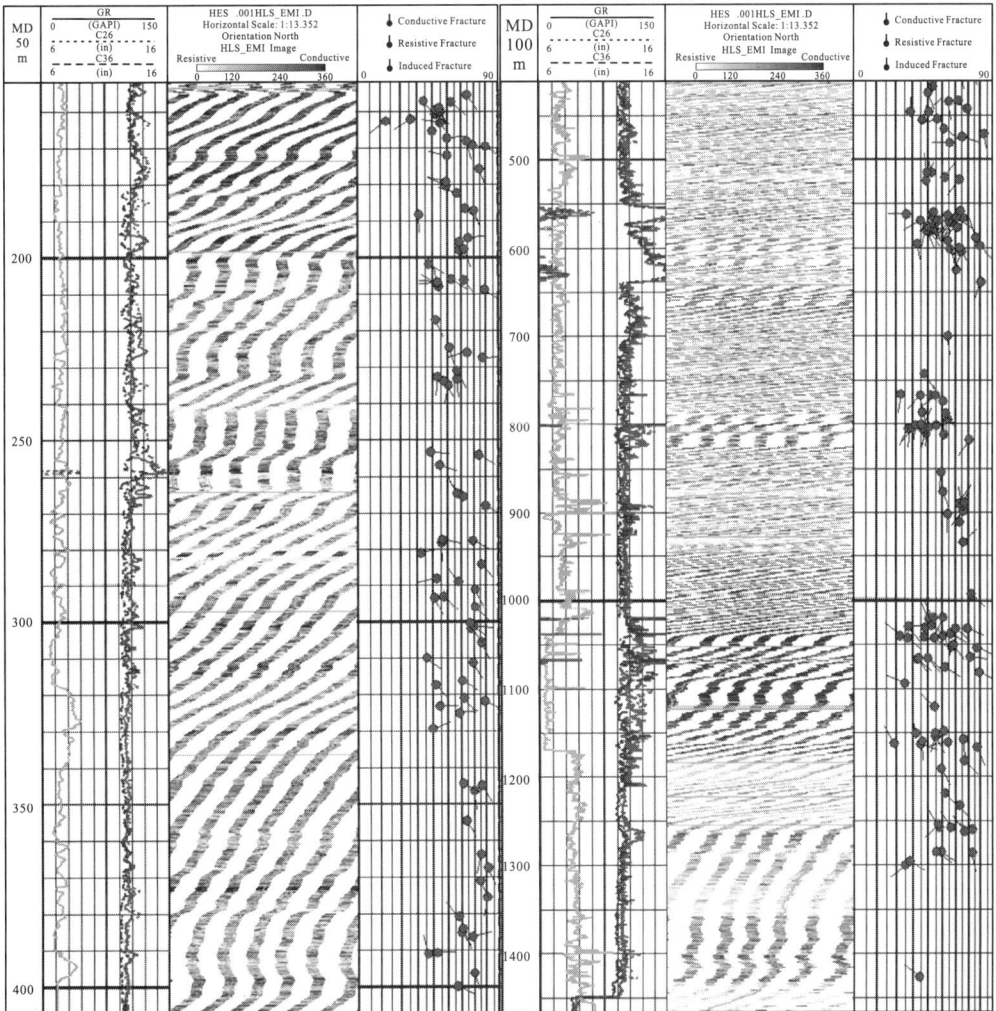

Figure 4-15　Calculation result of fracture parameters in Well HS2 (152-408 m, 408-1466 m).

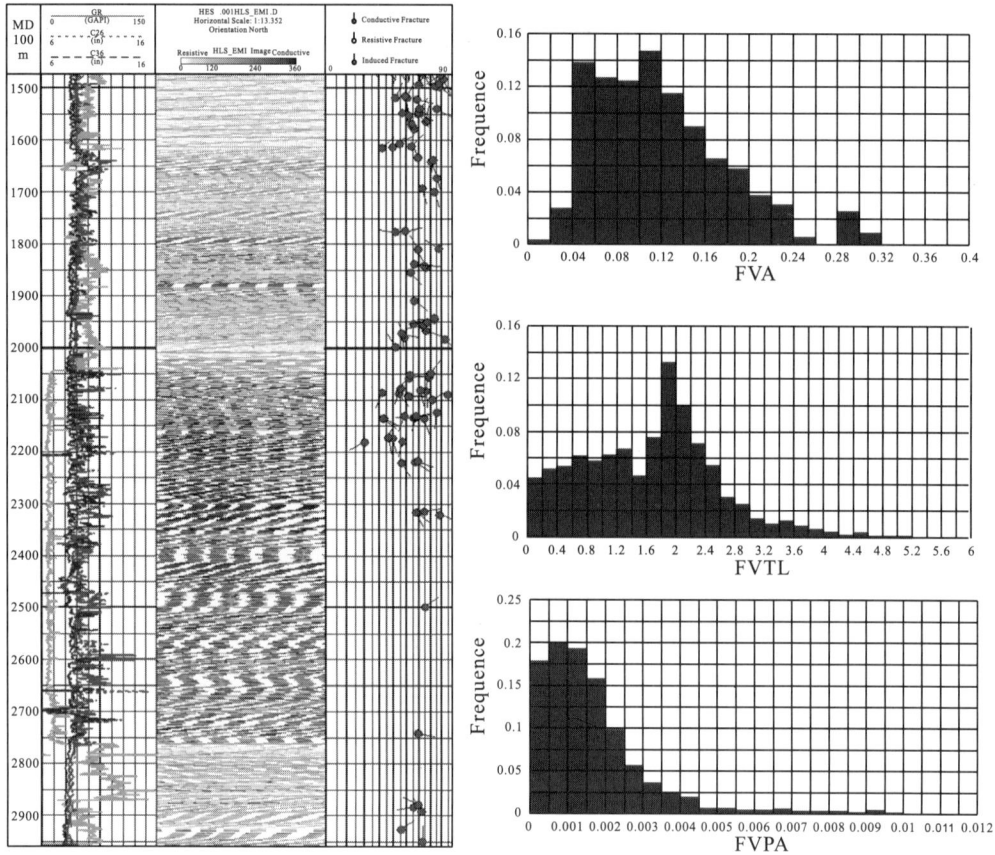

Figure 4-16 Calculated fracture parameters and statistics of fracture parameters in whole well section of Well HS2 (1466-2959 m).

3. Well HQ6

1) Development features of fractures

Fractures in the XRMI image of Well HQ6 are mainly high-conductivity fractures, high-resistivity fractures and induced fractures.

Generated during the drilling process and in feather shape, drilling induced fractures feature symmetric occurrence along borehole wall (Figure 4-17). The trend of induced fractures can better reflect the direction of current maximum horizontal stress. The induced fractures in this well are 60°-80° in dip angle, in NWW dip and NNE-SSW strike.

The high-conductivity fractures in electronic imaging pictures are shown as low-resistivity dark sine strip curves with usually dissolution along fractures, which are denoted by red tadpoles on the result map (Figure 4-18). The statistics of fracture occurrence indicates that high-conductivity fractures in this well are mainly middle-high angle fractures, with a dip angle of 20°-80°, trend of NW-SE and strike of NE-SW.

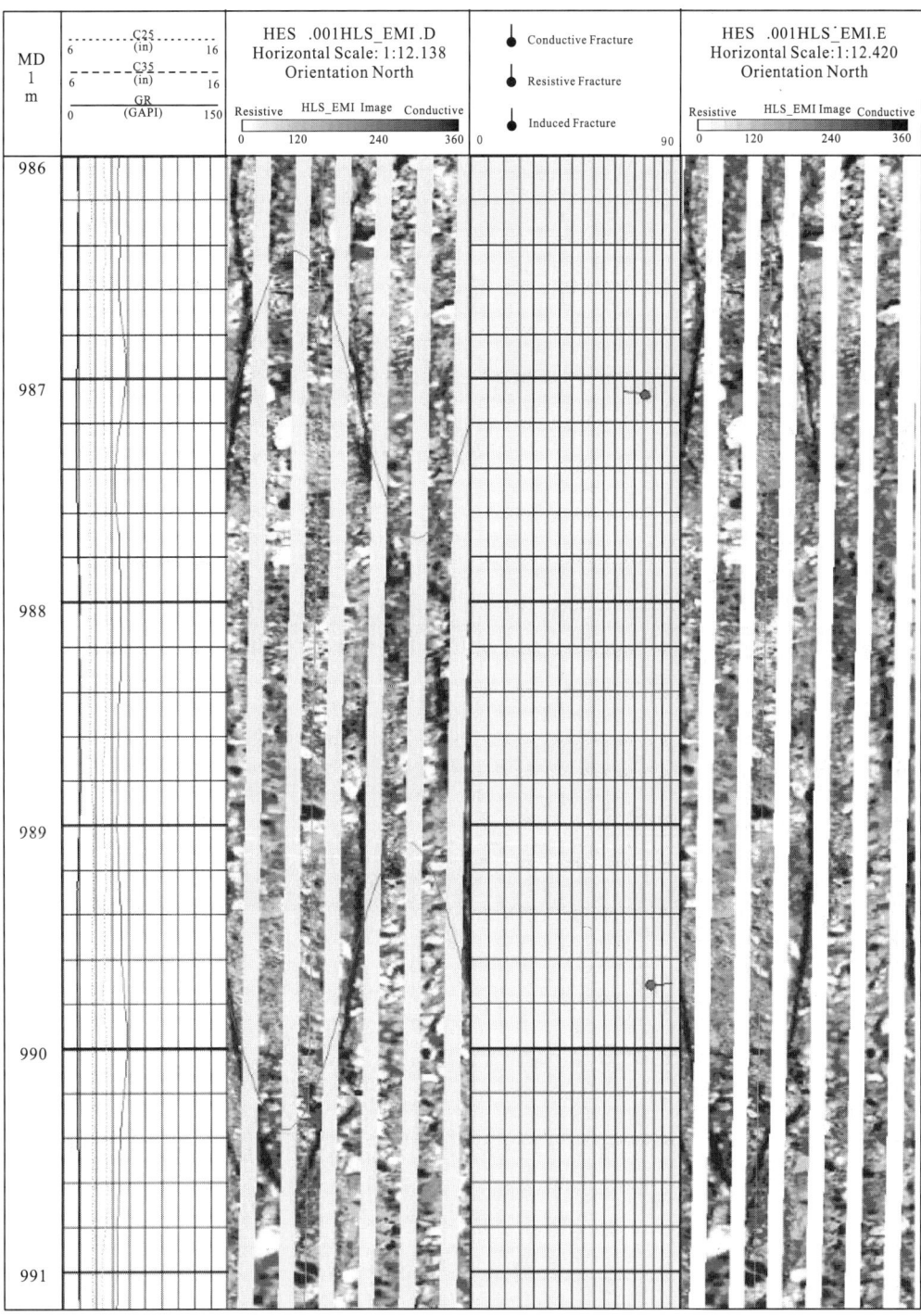

Figure 4-17 Features and occurrence statistics of induced fractures in XRMI image, Well HQ6 (987–991 m).

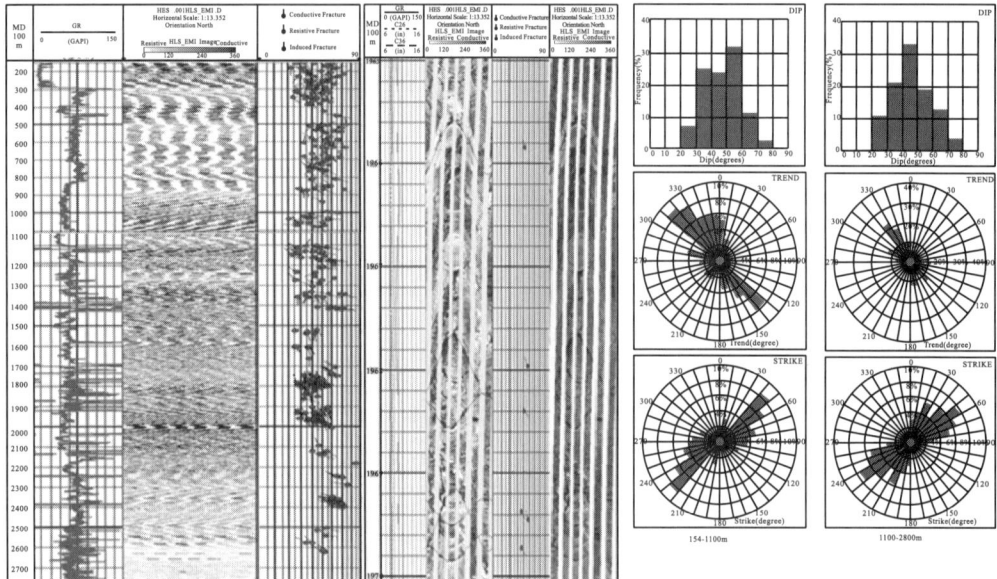

Figure 4-18 Statistics on imaging features, distribution and occurrence of high-conductivity fractures in Well HQ6 (154-2800 m) on XRMI image.

The high-resistivity fractures, closed or filled, are shown as bright yellow or white sine lines on the result map, and usually early stage fractures filled with cements or closed in the late stage (Figure 4-19). The occurrence statistics show that the high-resistivity fractures are 0°-90° in dip angle, and random in trend.

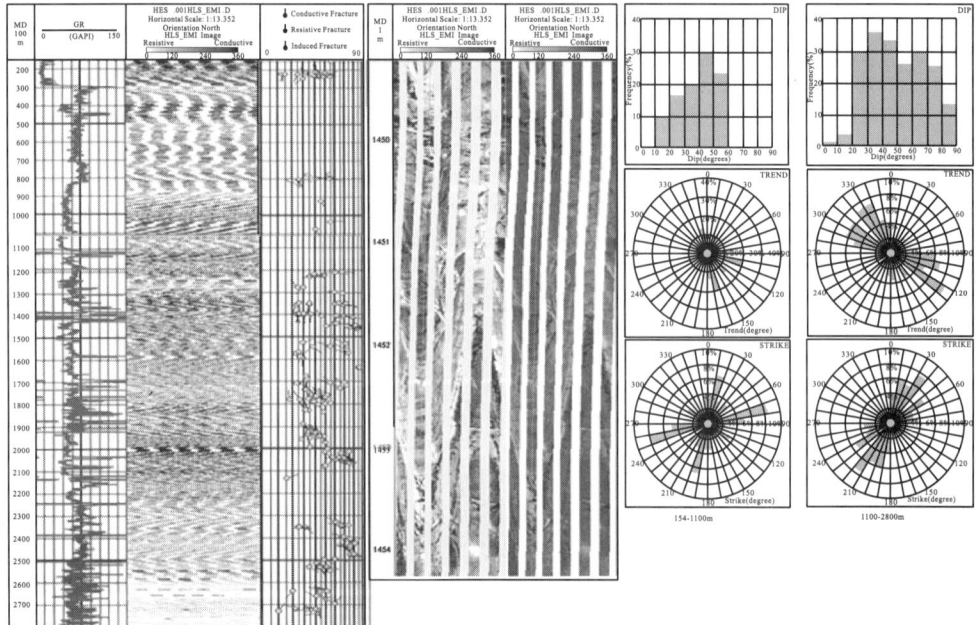

Figure 4-19 Statistics on imaging features, distribution and occurrence of high-resistivity fractures in Well HQ6 (154-2800 m) on XRMI image.

2) Quantitative calculation of high-conductivity facture parameters

The quantitative calculation results of fracture parameters (Figure 4-20) show that the fractures in Well HQ6 have a length <6 m/m^2, on average 2.35 m/m^2; hydrodynamic width <250 μm, on average 100 μm; and apparent porosity <0.006%, on average 0.001%.

The tuff section in Carboniferous of Well HQ6 is abundant in reservoirs, which are mainly pore, fracture-pore and dissolution-pore types, with dissolution gradually increasing from top to bottom, and rich secondary pores.

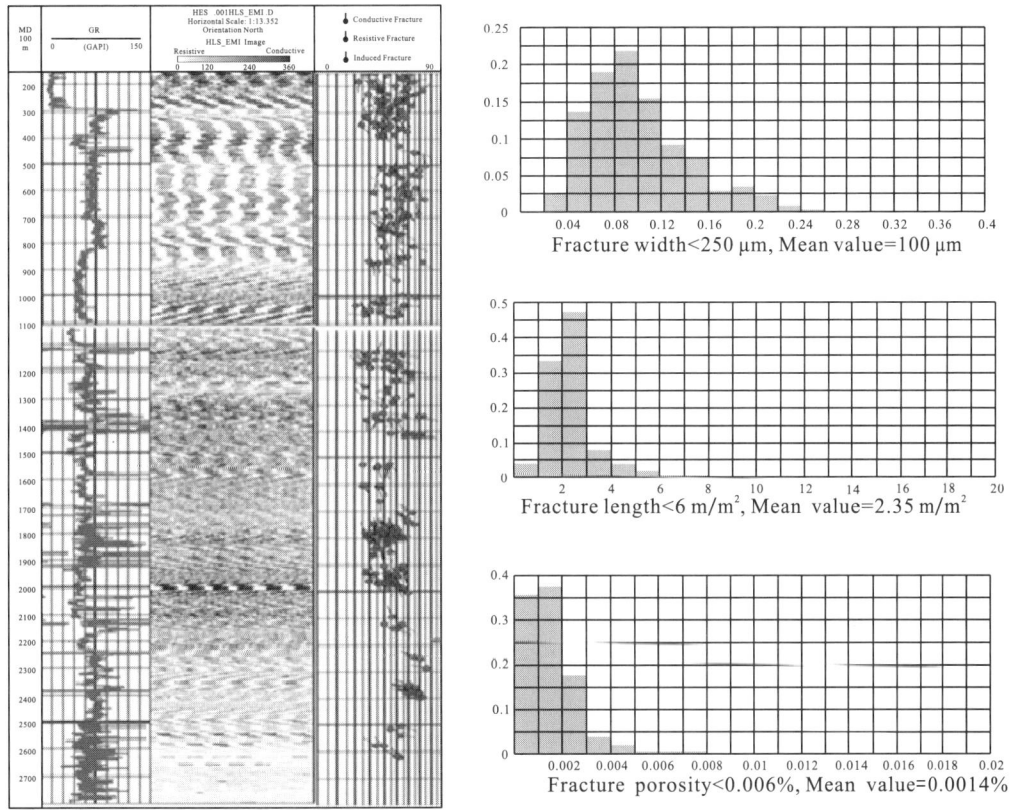

Figure 4-20 Statistics of fracture parameters in Well HQ6 (154–2800m).

4.2.2 High-conductivity fractures and high-resistivity fractures

Since natural structural fractures in Carboniferous and Permian are the interest in this study, the high-conductivity fractures and high-resistivity fractures on imaging logging pictures were counted. Classified according to different conductivity reflected during imaging logging, high-conductivity fractures represent fractures with high conductivity, and implicitly low filling degree and strong conduction capacity of fluid; while high-resistivity fractures represent fractures with low conductivity, implicitly high filling degree and poor or basically no conduction capacity for fluid.

Based on statistics of number and occurrence of high-conductivity fractures and high-resistivity fractures in the above five wells, the overall density and dominant orientation of structural fractures in Carboniferous, Permian in the Hala'alate Mountain area have been analyzed. The statistics of high-conductivity fractures, high-resistivity fractures and fracture density calculated from total fracture number (Table 4-2) can give us an insight into the overall development degree of fractures in Carboniferous and Permian of the Hala'alate Mountain area.

Table 4-2 Statistics of fracture density of Carboniferous–Permian in the Hala'alate Mountain area from imaging logging.

Well Name	Depth /m	Density of high conductivity fractures/m^{-1}	Density of high-resistivity fractures/m^{-1}	Overall fracture density/m^{-1}
HS1	1335.88–2552	0.40	0.06	0.46
HS2	1454.1–2957.8	0.45	0.06	0.50
HQ3	1501–2867.4	0.12	0.02	0.14
HQ6	154.2–2696.3	0.12	0.06	0.18
HQ101	1103.7–2310.2	0.13	0.03	0.16

1. High-conductivity fractures

1) Dominant orientation of high-conductivity fractures

Based on statistics of all high-conductivity fractures in imaging logging sections in Well HS1, HS2, HQ3, HQ6 and HQ101, the rose map of high-conductivity fracture dip of related wells have been made (Figure 4-21). Among them, the dominant orientation of high conductivity fractures in Well HS1, HS2 and HQ101 is SE; and that in Well HQ3 and HQ6 is NW. In general, the dominant dip orientation of high-conductivity fractures picked from imaging logging in the Hala'alate Mountain area is NW and SE.

2) Distribution features of high-conductivity fracture dip angles

To facilitate the analysis of distribution features of high-conductivity fracture dip angles, high conductivity fracture dip angles of logging section in the above five wells have been counted up, and frequency histogram of high conductivity fracture dip angles of the five wells have been made (Figure 4-22), which can clearly reflect the main range of fracture dip angle in each well. The high conductivity fractures in Well HS1, HQ3, and HQ101 mainly are mostly high angle fractures, with the dip angle range peaking respectively at 60°–70°, 70°–80° and 50°–60°. In contrast, high conductivity fractures in Well HS2 and HQ6 have no obviously dominant dip angle, with wider dip angle range and smaller dip angle peak of mainly 30°–40°. It can be seen from the distribution range of dip angle of high-conductivity fractures in the five wells, the most concentrated range is 50°–70°, and the high conductivity fractures in the area are more likely to be high-angle on the whole.

Chapter 4 Development Features of Reservoir Fractures

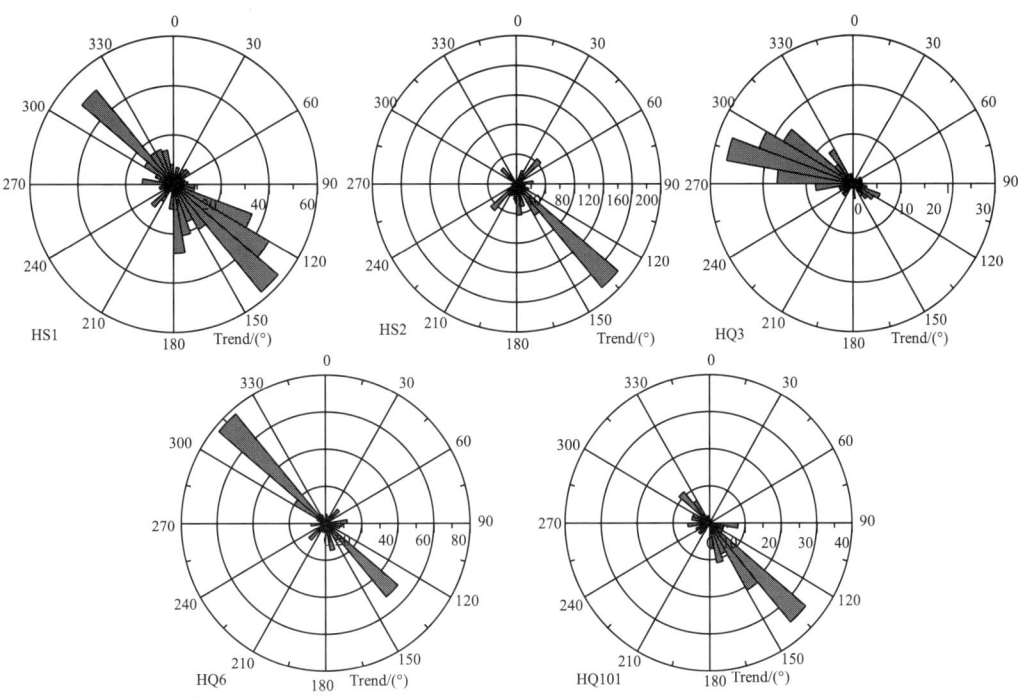

Figure 4-21 Rose map of trend of Carboniferous-Permian high-conductivity fractures in five wells of the Hala'alate Mountain area.

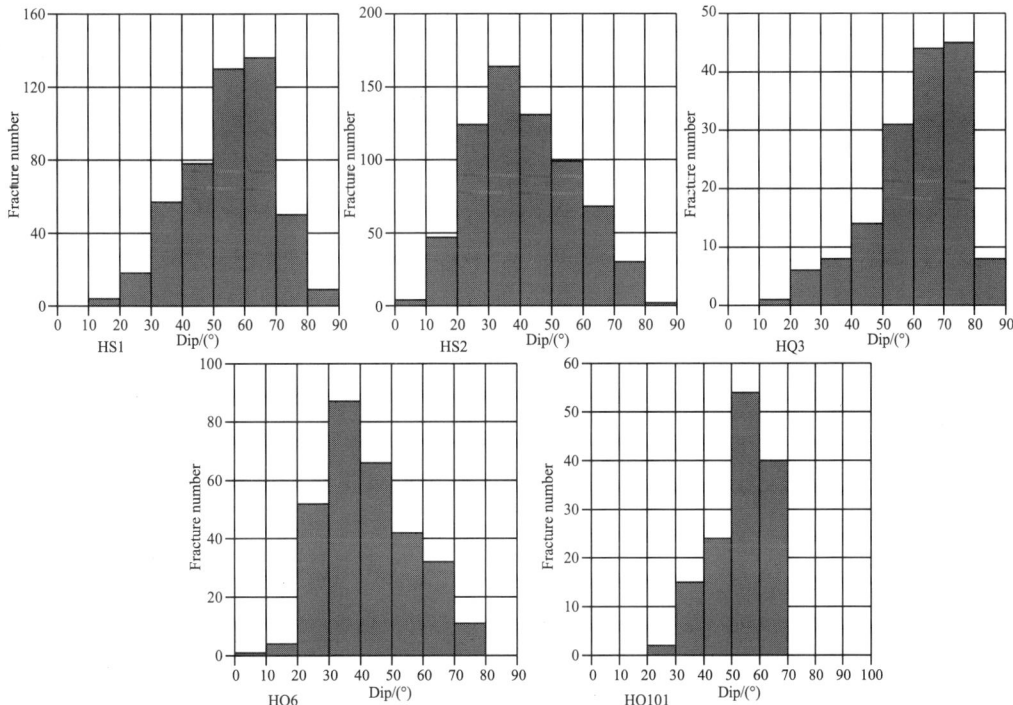

Figure 4-22 Histogram of dip angle of high-conductivity fractures in Carboniferous-Permian of the Hala'alate Mountain area.

3) Development density of high-conductivity fractures

The development degree of high-conductivity fractures in strata directly affects their contribution to oil and gas migration and accumulation, so it is necessary to conduct statistics and analysis on fracture development density of the logging sections. The histogram of high-conductivity fracture density of the five wells show (Figure 4-23) that Well HS1 and HS2 have a higher fracture density of $0.4/m^{-1}$ and $0.45/m^{-1}$ respectively, while Well HQ3, HQ6 and HQ101 have a lower fracture density of $0.12/m^{-1}$, $0.12/m^{-1}$ and $0.13/m^{-1}$ respectively. The obvious difference in fracture density is clearly closely related to the structural location of the wells, in other words, structural fractures are more developed in the location with strong structural activity or tectonic stress. The statistics and comparison of high-conductivity fracture density show that compared with Well HQ3, HQ6 and HQ101, Well HS1 and HS2 are located in the structural locations with more complex geological conditions.

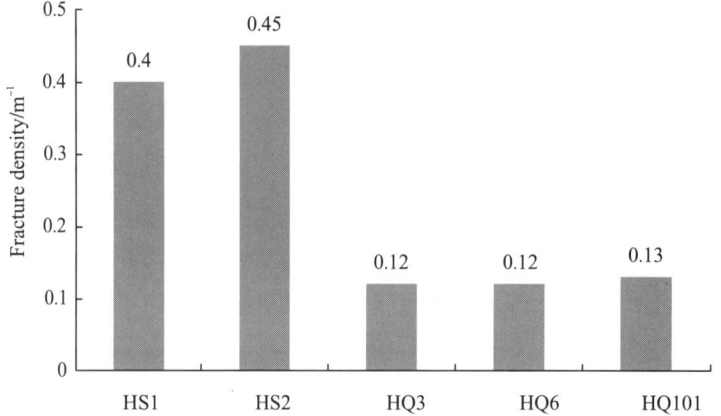

Figure 4-23 Histogram of high-conductivity fracture density in the research area.

2. High-resistivity fractures

1) Dominant orientation of high-resistivity fractures

High-resistivity fractures are another type of natural structural fractures opposite to high-conductivity fractures in imaging logging, which have higher degree of filling, and contribute less to formation fluid seepage and conduction. Based on statistics of high resistivity fracture and in imaging logging pictures of the Hala'alate Mountain area, the rose map of high resistivity fracture trend of the five wells have been made (Figure 4-24). The trend rose map of the five wells show that the dominant trend of high-resistivity fractures in Well HS1, HS2 and HQ6 is SE, that of Well HQ3 is SW and NW and that of Well HQ101 is SE, NW and NE, mainly SE and NW.

2) Distribution features of high-resistivity fracture dip angle

High resistivity fractures are obviously different from high conductivity fractures in terms of conduction to formation fluid. In order to analyze their difference, dip angle of high-resistivity fractures in the imaging logging sections of five wells (Well HS1, HS2, HQ3, HQ6 and

HQ101) have been counted up, and histogram of the five wells have been made (Figure 4-25).

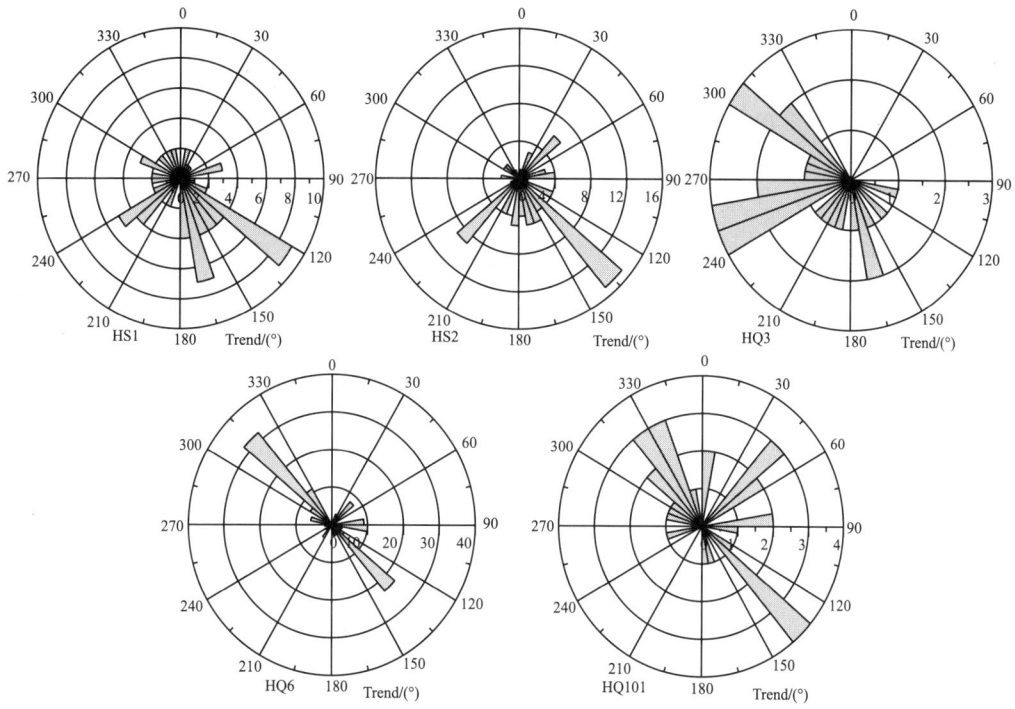

Figure 4-24 Trend rose map of high-resistivity fractures in Carboniferous-Permian of the Hala'alate Mountain area.

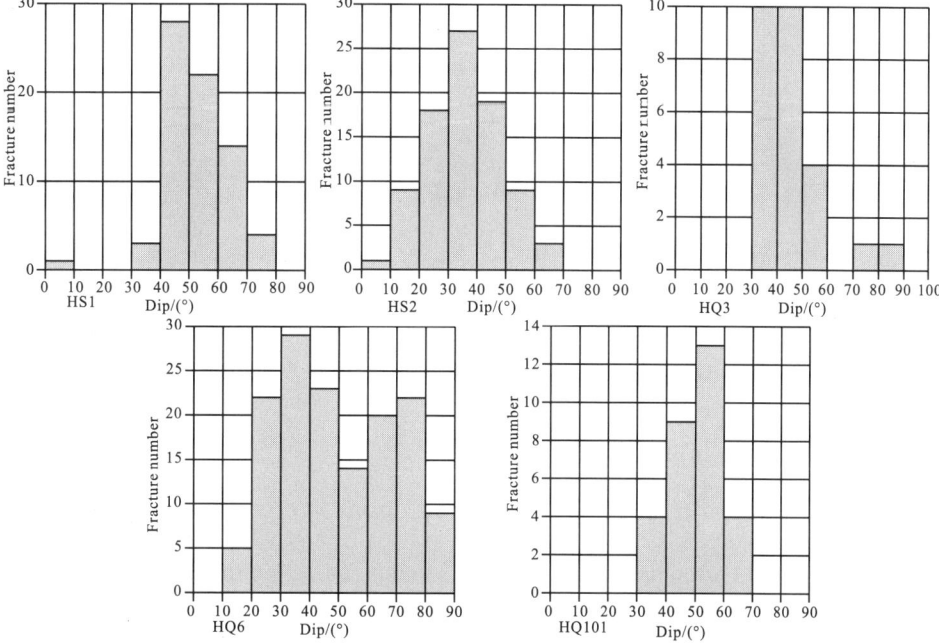

Figure 4-25 Histogram of dip angle of high-resistivity fractures in Carboniferous-Permian of the Hala'alate Mountain area.

The high resistivity fractures in Well HS1 mianly concentrate in the dip angle of 40-60°, with peak of 40°–50°. The high resistivity fractures in Well HS2 and HQ3 concentrate mainly in the dip angle of 30°–50°, with peak of 30°–40°. The high resistivity fractures in Well HQ6 have no concentrated dip angle, with a large dip angle range of 10°–90°, there are two peaks in the histogram: 30°–40° and 60°–70°, with the former being the dominant peak. The dip angle of high resistivity fractures in Well HQ101 concentrate mainly in 40°–60°, with the peak of 50°–60°.

The high-resistivity fractures in imaging logging sections of the Hala'alate Mountain area concentrate in the dip angle of 30°–60°, representing middle-angle fractures, a transitional type between low-angle fractures and high-angle fractures.

3) Development density of high-resistivity fractures

The statistical results of fractures from imaging logging (Figure 4-26) show that Well HQ101 and HQ3 have much higher density of high resistivity fractures than Well HS1, HS2 and HQ3. But for the whole research area, the development density of high resistivity fractures is far lower than that of high conductivity fractures, which suggests that structural fractures in Carboniferous-Permian in the Hala'alate Mountain area have stronger conduction capacity for fluid.

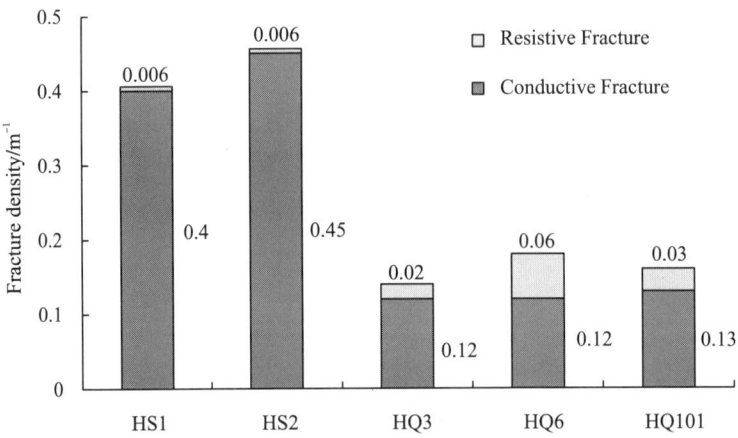

Figure 4-26 Histogram of development density of high resistivity fractures and high conductivity fractures in the Hala'alate Mountain area.

3. Occurrence features of fracture development sections in typical wells

The Carboniferous and Permian in the Hala'alate Mountain area are tight, and the development of structural fractures is of vital importance to the formation of favorable reservoir, therefore, while conducting statistics of the development conditions of high conductivity fractures and high resistivity fractures by using imaging logging data, the concentrated sections of fractures have also been examined, which shows that each well has concentrated sections of structural fractures.

1) Well HS1

The distribution graph of fractures in the logging section of Well HS1 (Figure 4-27a) shows that fractures are developed across the whole imaging logging section and more developed in two sections, 1600–1800m and 2400–2600m, and the two sections are both andesite of Carboniferous.

The fractures in the shallower 1600–1800m interval of Well HS1 are mainly 50°–60° in dip angle (Figure 4-27b), and SE in dominant orientation (Figure 4-27d); while fractures in the 2400–2600m interval are 60°–70° in dip angle (Figure 4-27c), and NW in dominant orientation (Figure 4-27e). The fractures in the two well sections are quite different in occurrence, which indicates that the fractures in different depths are different in orgin to some extent.

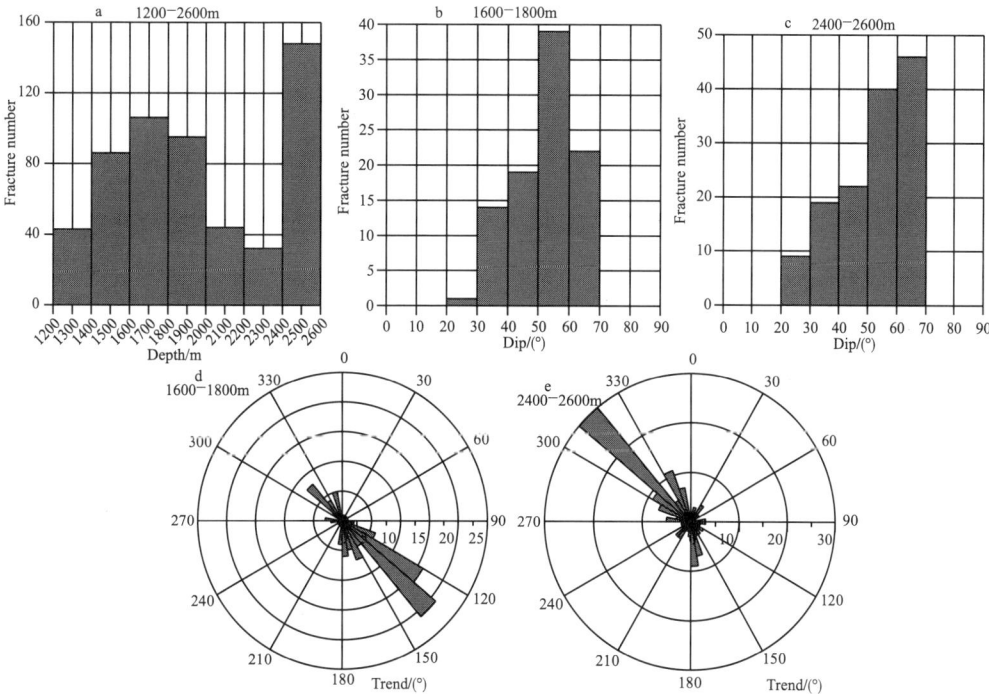

Figure 4-27　Occurrence statistics of fractures in the concentrated section of Well HS1 from imaging logging.

2) Well HS2

The distribution graph of fractures in the logging section of Well HS2 shows that fractures are quite developed in the whole logging section, especially at the depth of 1400–1600m (Figure 4-28a). The fracture concentrated section is andesite of Carboniferous, where the fractures are mainly high-angle ones, with dip angle of mainly 50°–60° (Figure 4-28b), and dominant orientation of SE (Figure 4-28c). This well is quite developed with fractures, implying that the location of the well has undergone stronger tectonic deformation.

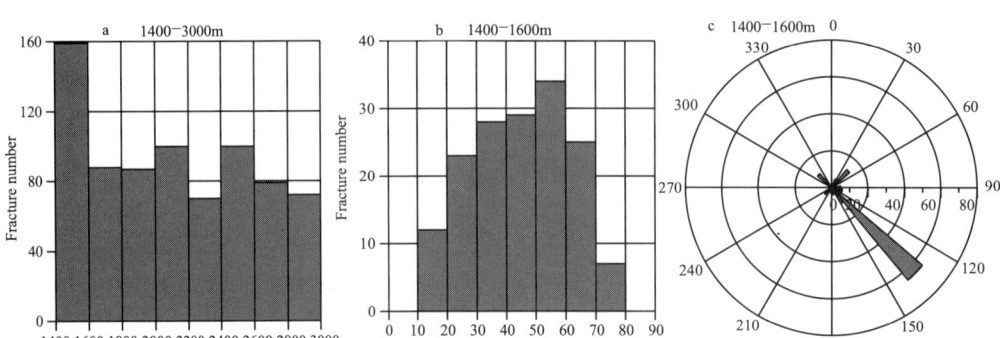

Figure 4-28 Occurrence statistics of fractures in the concentrated section of Well HS2 from imaging logging.

3) Well HQ3

The distribution graph of fractures in the logging section of Well HQ3 (Figure 4-29a) shows that fracture density differs widely in different depths. The fractures in 2000–2200m are much higher in density. The fractures are mainly high-angle fractures in volcanic breccia of Carboniferous, 60°–80° in dip angle (Figure 4-29b), and NWW in trend (Figure 4-29c).

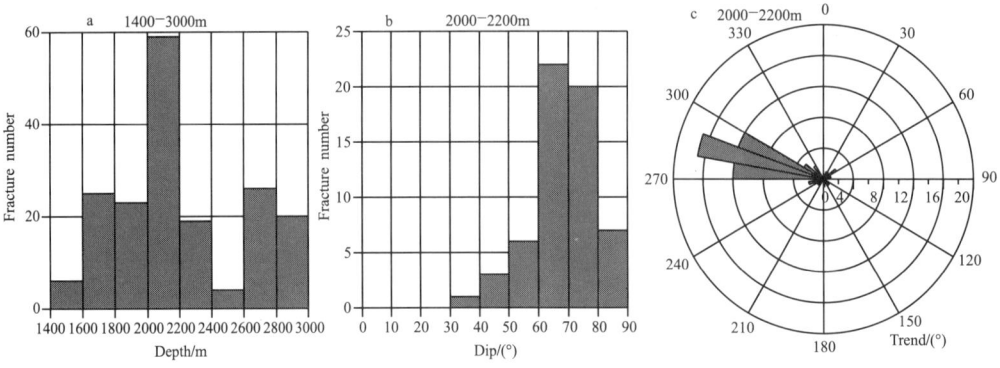

Figure 4-29 Occurrence statistics of fractures in concentrated section of Well HQ3 from imaging logging.

4) Well HQ6

Fractures in Well HQ6 have higher development degree in 800–2000m, and are concentrated in 1600–2000m mudstone section of Permian. The fractures in this section have a smaller dip angle of 30°–50° (Figure 4-30) and the dip angle peak of 30°–40° (Figure 4-30). The fractures in the concentrated section basically have two trends, NW and SE, with the latter being the dominant orientation, which indicates that the fracture concentrated section of Well HQ6 is an overlay area of multi-phase tectonic activities.

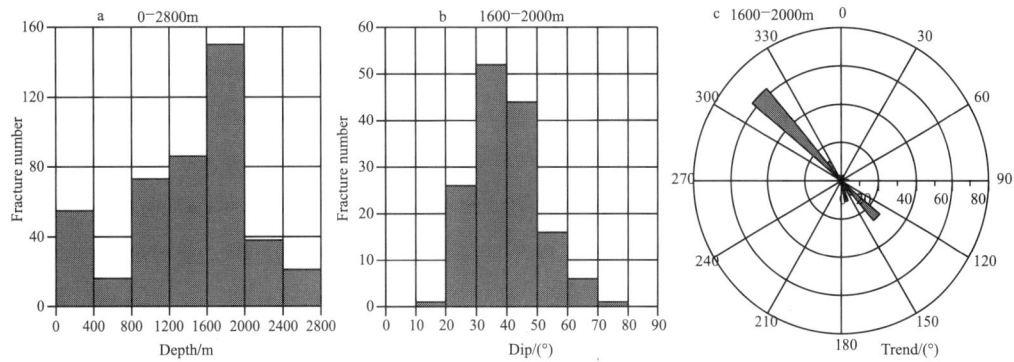

Figure 4-30 Occurrence statistics of fractures in the concentrated section of Well HQ6 from imaging logging.

5) Well HQ101

The fracture distribution graph of logging section in Well HQ101 (Figure 4-31a) shows that fractures are more developed in 1600–1800m and 2000–2200m (more developed in the latter section), and the two sections are volcanic breccia of Carboniferous and Permian mudstone respectively.

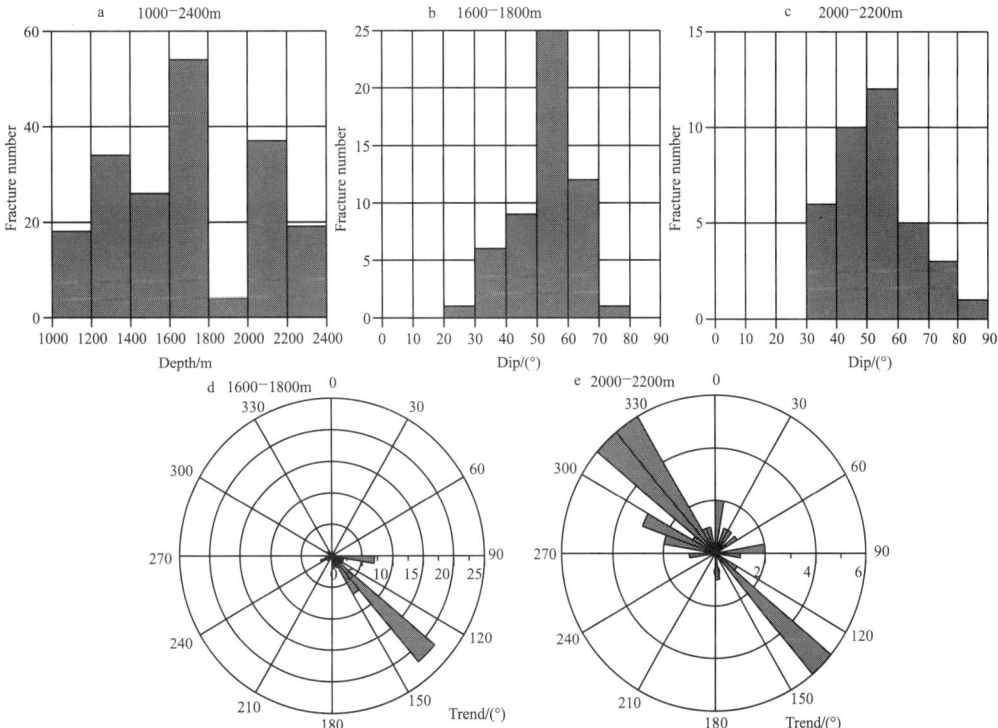

Figure 4-31 Occurrence statistics of fractures in the concentrated section of Well HQ101 from imaging logging.

Based on the analysis of fracture occurrence in two concentrated sections of Well HQ101, it is found that fractures in Carboniferous volcanic breccia of 1600–1800 m mainly concentrate in the dip angle of 50°–60° (Figure 4-31b), and orientation of SE (Figure 4-31d); while fractures in the Permian mudstone of 2000–2200m are mainly in the dip angle of 40°–60° with the peak dip angle of 50°–60° (Figure 4-31c), and mainly in NW and SE orientation, with the latter accounting for a larger ratio (Figure 4-31e). This indicates that the fracture concentrated section of 2000–2200 m in Well HQ101 has similar overlay features of multi-phase structural fractures with that of 1600–2000 m in Well HQ6.

Chapter 5

Geochemistry of Fracture Fillings

Fractures in rock are important pathways for fluid flow in a basin. In the process of fluid migration along fractures, minerals in the fluid tend to crystallize, precipitate and form fracture fillings due to the variations in temperature, pressure and salinity (Morrow, 1982). The geochemical information implied in these fillings is of great importance for the study of fluid activity pattern and the hydrocarbon accumulation process. Although there are many different kinds of fillings (such as quartz, clay, and carbonate etc), the most common minerals in reservoirs are carbonate minerals. As this kind of mineral is sensitive to medium environment, able to better record multiphase fluid activity process and easy to test, analyze and compare, they have become ideal objects for geologists to conduct reservoir geochemical analysis (Hu et al., 2009). In recent years, with the progress of geochemical analysis methods, geochemical analysis of carbonate minerals in reservoir has become an important way for petroleum geologists to study fluid activity pattern and hydrocarbon accumulation process in basins, and good results have been achieved in a series of studies(Liu et al., 2004; Wang and Zhang, 2001). Common geochemical analysis means of carbonate minerals at present include analysis of carbon, oxygen, and strontium isotopes, and analysis of fluid inclusion, etc.

5.1 Isotope geochemistry

5.1.1 Sr isotope

Strontium is a kind of trace element widely distributed in nature. Its atomic structure, crystal chemical characteristics and geochemical behavior are similar to that of calcium. Strontium often replaces calcium and enters into carbonate crystal lattice with the isomorphism, but once entering, it won't have isotope fractionation, and thus its isotopic composition at the time of fluid deposition will be reserved. If a mineral is not strongly transformed by fluid of other nature in later stage, its strontium isotope composition is only controlled by three main original input mechanisms (mantle source strontium, crust source strontium and re-dissolved strontium) (Banner and Jay, 2004; Palmer and Edmond, 1989; Palmer, 1985). Today the strontium isotope composition obtained in the test of carbonate mineral can basically represent the strontium isotope composition of original geological fluid (Wang and Liu, 2009). Therefore, comparison of strontium isotope composition can be used to trace the fluid source that forms the carbonate minerals.

In recent years, strontium isotope research method has gradually become a new geochemical research tool of sedimentary isotope. Strontium isotope geochemical analysis has achieved good application effect in the study of stratigraphy, sedimentology, diagenesis and hydrocarbon accumulation, etc. (Yuan et al., 2014; Liu, 2007; Huang et al., 2002). Calcite is widespread in the volcanic rock fractures in the study area, laying a good material foundation for the geochemical study of strontium isotope. Due to the lack of isotope geochemical study in the study area, strontium isotope geochemical analysis of the calcite vein in the study area is of great significance for telling the stages and nature of ancient geological fluid in the volcanic reservoirs.

The strontium isotope composition in carbonate cement is mainly influenced by fluid source. The common fluid sources include: seawater, lake water, and diagenetic fluid, etc. Now it is generally believed that the strontium isotope in seawater or rock is mainly controlled by the crust source and mantle source strontium. The crust source strontium mainly comes from the ancient land rock weathering, thus the crust source material is rich in Rb and high in $^{87}Sr/^{86}Sr$ ratio, and the average $^{87}Sr/^{86}Sr$ ratio of the global crust source strontium is 0.7119. Mantle source strontium mainly comes from hydrothermal fluid system of the mid-ocean ridge. Mantle source materials have extremely low content of Rb, and lower $^{87}Sr/^{86}Sr$ ratio, and the average $^{87}Sr/^{86}Sr$ ratio of the global mantle source strontium is 0.7035 (Palmer and Edmond, 1989; Palmer, 1985). As the strontium can stay in seawater for as long as one million years and the mixing time of the seawater is only one thousand years, the strontium isotope composition of the global seawater in the same period of geological history should be theoretically uniform. Therefore, $^{87}Sr/^{86}Sr$ ratio can be used to distinguish different fluid sources (Veizer et al., 1997).

In order to find out the formation mechanism and development stages of fracture fillings, and the relationship between fracture development and hydrocarbon accumulation in Carboniferous-Permian volcanic rock in the Hala'alate Mountain area in the northwestern margin of Junggar Basin, calcite filling in structural fractures from the cored sections of 5 representative wells in the area has been sampled, and strontium isotope analysis has been conducted on them in the research (Table 5-1).

Chapter 5 Geochemistry of Fracture Fillings

Table 5-1 Analysis data of strontium isotope of the calcite vein in fractures in the study area.

Sample No.	Depth/m	Horizon	Description	$^{87}Sr/^{86}Sr$
HQ101-1	1455.12	C	Volcanic breccia, with calcite filling in the fractures and oil immersion on the section	0.707064
HQ101-3	1739.00	C	Volcanic breccia, with calcite filling in fractures and oil immersion on the section and crude oil in the fractures	0.705634
HQ101-5	1739.95	C	Volcanic breccia, with calcite filling in fractures and oil immersion on the section and crude oil in the fractures	0.707415
HQ101-6	1740.88	C	Volcanic breccia, with calcite filling in fractures and bead-like distribution of oil patches	0.705542
HQ101-7	2237	P	Mudstone, with calcite filling in fractures and crude oil in the dissolution part of the calcite	0.705124
HQ101-9-2	2240.7	P	Mudstone, with calcite filling in fractures and crude oil in the dissolution part of the calcite	0.705407
HQ3-2-2	1475.92	C	Volcanic breccia, with calcite filling in fractures and bitumen on the section	0.705836
HQ3-3	1478.19	C	Volcanic breccia, with calcite filling in fracture and bitumen on the section	0.705488
HQ4-2	686.40	C	Tuff, with calcite and quartz filling in fractures	0.703264
HQ4-3	686.50	C	Tuff, with calcite and quartz filling in fractures	0.703292
HQ6-2	255.50	C	Tuff, with calcite and quartz filling in fractures	0.703345
HQ6-3	256.00	C	Tuff, with calcite filling in fractures; fine pyrite particles distributed along the fractures	0.703412
HQ7-1	162.27	C	Tuff, with calcite filling in fractures	0.70527
HQ7-2	211.66	C	Tuff, with calcite filling in fractures	0.704797
HQ7-3a	212.00	C	Tuff, with calcite filling in fractures	0.704902

The strontium isotope test results show that $^{87}Sr/^{86}Sr$ values of all the Carboniferous-Permian fracture-filling calcite in the study area are less than that of modern seawater (with an average $^{87}Sr/^{86}Sr$ value of 0.709073), and the value of some samples are even less than the lowest evolution $^{87}Sr/^{86}Sr$ value (0.707) since the phanerozoic in seawater (Hoefs, 1997; Koch et al., 1992), which indicates that isotopic composition of the fracture-filling carbonate minerals is affected by mantle source strontium. $^{87}Sr/^{86}Sr$ ratio of the most test samples from the Hala'alate Mountain area are higher than the mantle source strontium boundary (0.7035), but apparently lower than crust source strontium standard (average $^{87}Sr/^{86}Sr$ of 0.7119), suggesting that most of the fracture-filling carbonate minerals in the Hala'alate Mountain area are affected by mantle source strontium and crust source strontium jointly. According to the test data of strontium isotope and the sampling depth, a depth distribution diagram of $^{87}Sr/^{86}Sr$ isotope ratio has been drawn (Figure 5-1). The strontium isotope ratio of the tested samples of calcite vein mainly concentrate in three ranges (0.703264–0.703412, 0.704902–0.705836, and 0.707064–0.707415), which respectively represent calcite of three different genetic types.

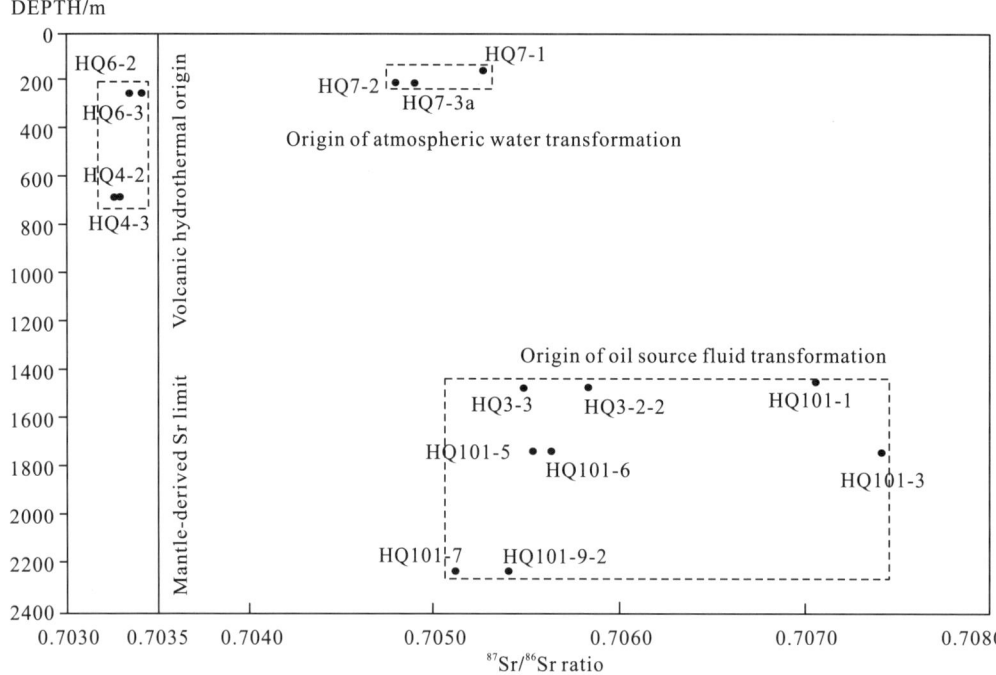

Figure 5-1 Relationship between the isotope ratio $^{87}Sr/^{86}Sr$ and depth of the Carboniferous-Permian calcite vein in the study area (mantle source strontium boundary value from Palmer and Edmond, 1989; Palmer and Elderfield, 1985).

The first type is the volcanic hydrothermal calcite: strontium isotope composition of the five samples with the $^{87}Sr/^{86}Sr$ ratio between 0.703264 and 0.703412 all falls into the mantle source strontium (average $^{87}Sr/^{86}Sr$ ratio of 0.7035) range, moreover, the cores of these samples has no oil and gas show, the calcite is high in crystal degree, coarse in crystal particles and pure in surface. The gas-liquid two phase brine inclusions captured in the calcite have generally higher homogenization temperature. Thin section observation shows there are signs of late volcanic hydrothermal process, namely, holes and fractures successively filled with zeolite and calcite etc in Carboniferous of Hala' alate Mountain area (Figure 5-2). According to previous studies, volcanic hydrothermal activities occurred in the Northwest Margin of Junggar Basin in the depositional stage of Permian Fengcheng Fm, causing the development of hydrothermal minerals such as zeolite and calcite in the fractures (Shi et al., 2013). Water-rock reaction would occur between the volcanic hydrothermal fluid and pre-existed igneous rock in the process of hydrothermal fluid activities, and thus volcanic hydrothermal fluid will obtain poor radioactive mantle source strontium from igneous rock, making the volcanic hydrothermal calcite has the $^{87}Sr/^{86}Sr$ ratio of mantle source strontium. Therefore, the study shows that the above 5 samples are of volcanic hydrothermal genesis.

Figure 5-2 Volcanic fractures successively filled with zeolit and calcitTuff, Carboniferous, Well HQ7, 211.60 m (a: Plane-polarized light; b: cross-polarized light).

The second type is the calcite formed by superimposed transformation of atmospheric water: the 3 samples from Well HQ7 in the structural high in northwest the Hala'alate Mountain area have a $^{87}Sr/^{86}Sr$ ratio between 0.704902 and 0.705836, higher than that of volcanic hydrothermal calcite. The study shows that the root cause of the above strontium isotope characteristic is that the calcite had crust source strontium and mantle source strontium (dominant) in the process of formation As the Hala'alate Mountain area is not big, the calcite with poor radioactive strontium formed after the early Permian volcanic hydrothermal activities should be widely distributed in the study area, and if the early volcanic hydrothermal calcite with the characteristic of mantle source strontium was subject to superposed transformation of later crust source strontium fluid, the $^{87}Sr/^{86}Sr$ ratio of calcite would rise to a certain extent. The 3 tested samples from Well HQ7 in this study are all shallow in burial depth (about 200 m). Iron oxide traces can be observed clearly on the cores where the calcite samples were taken (Figure 5-3a, Figure 5-3b), which clearly shows that the calcite filling the fractures early has subjected to superposed transformation of later oxygen-enriched atmosphere water infiltrating along the fractures. In the process of seeping down, atmospheric water dissolved aluminum silicate minerals in the Carboniferous igneous rock, releasing radioactive strontium. The atmospheric water containing radioactive strontium would undoubtedly lead to the rise of $^{87}Sr/^{86}Sr$ ratio of the calcite after the superposed transformation. Since no oil and gas shows have been found in these cores and thin sections (Figure 5-3a, Figure 5-3b), the rise in $^{87}Sr/^{86}Sr$ ratio is probably not related to oil and gas. Therefore, strontium isotope characteristics of the samples from Well HQ7 should be mainly related to the superposed transformation of atmospheric water.

The third type is the calcite caused by oil source fluid: the 9 samples with $^{87}Sr/^{86}Sr$ ratio between 0.707064 and 0.707415 are similar in strontium isotope characteristic to the calcite sample caused by superposed transformation of atmospheric water. The strontium isotope of these calcite samples show the feature of mixed crust source strontium and mantle source strontium. Observation of core section of the above samples shows that the coarse-grain calcite filling in the fractures has late dissolution and the dissolution parts all have traces of crude oil filling (Figure 5-3c, Figure 5-3d). Abundant secondary hydrocarbon inclusions along the calcite fractures can be observed in the inclusion slices. The study shows that the above coarse crystalline calcite filling in the fracture has similar crystal shape with the hydrothermal calcite, so its initial deposition mechanism should be related to the volcanic hydrothermal process. In addition, some dissolution parts in the calcite vein filling in the core have crude oil attached, suggesting that the early precipitated calcite has subjected to the superposed transformation of hydrocarbon fluid related to the late hydrocarbon generation and expulsion of source rock. As the oil source of the Carboniferous igneous rock in the Hala'alate Mountain area mainly comes from Permian clastic formations (Shi et al., 2013), and formation fluid with rich radioactive strontium coming from the hydrocarbon generation and expulsion of source rock transformed the early volcanic hydrothermal calcite during its filling in the Hala'alate Mountain area, making the $^{87}Sr/^{86}Sr$ ratio of calcite after transformation higher than that of mantle source strontium, but lower than crust source strontium. Therefore, it is concluded from comprehensive analysis that the above 9 calcite samples are formed by oil source fluid transformation.

Figure 5-3 Typical characteristics of Carboniferous-Permian cores. a: Tuff, HQ7, Carboniferous, 210–213.15 m; b: Tuff, HQ7, Carboniferous, 211.6 m; c: Volcanic breccia, Well HQ101, Carboniferous, 1737.70 m; d: Mudstone, Well HS1, Permian, 2099 m.

5.1.2 C and O isotopes

The geological reactions concerned by petroleum geologists often occur in water, including mineral deposits from sea water and diagenesis that may occur in other types of water (Emery and Robinson, 1993). At present, the use of isotope tracing to study water types and action processes that form rock or minerals have gradually become a mainstream technology (Li, 1998; Criss et al., 1987; Craig, 1966).

Carbon and oxygen are important elements of carbonate, and geochemical characteristic analysis of carbon and oxygen isotope plays an important role in tracing fluid source, calculating mineral-forming temperature, and studying diagenetic environment, etc (Du et al., 2005), therefore, such analysis method is frequently used by geologists (Zhu et al., 2015; Gao et al., 2011) But as geochemical characteristics of carbon and oxygen isotopes are related to fluid source, diagenetic environment, and many other factors, they have multiple explanations, so the geochemical characteristics analysis of carbon and oxygen isotopes in the study area is used as supplement to the analysis of strontium isotope geochemical characteristics.

Volcanic rock fractures are filled with calcite veins in general, which provide material condition for using geochemical characteristics of C and O isotopes to find out carbon source, mineralization temperature, fracture stages, and hydrocarbon accumulation, etc. in the study area. The existing researches show that the carbon isotope in the study of diagenesis is mainly used to reveal the source of carbon in carbonate minerals (Clark and Fritz, 2000). In sedimentary basin, there are mainly two kinds of carbon reserves: ①from marine carbon or carbon formed by chemical precipitation; ②from reducing organic carbon (Emery and Robinson, 1993). These two kinds of carbon isotopes can be identified through their isotopic signature. $\delta^{13}C$ values of most marine facies sources are ±4‰ (Deines, 1980), $\delta^{13}C$ values of lake water carbonate rock are usually heavier (2.9‰–9.3‰) (Liu, 1998.). By contrast, $\delta^{13}C$ values of organic carbon generally range between −10‰ and −35‰, and the most typical $\delta^{13}C$ values are generally between −20‰ and −30‰. Oxygen isotope is usually taken as the geological thermometer for the mineral formation. The study on the Bohai Bay area by Wang Darui and Zhang Yinggong showed that under the influence of high temperature of diagenesis and metamorphism, $\delta^{18}O$ in the water would be consumed largely, leading to negative $\delta^{18}O$ value of rock formed. Therefore, $\delta^{18}O$ and $\delta^{13}C$ can be combined to analyze the formation conditions of fracture fillings. But since the factors that influence C and O isotopes are complex, the tested C and O isotope values may have many interpretations. Therefore, the C and O isotope tests must combine with other analysis and test results and fundamental geology to get reliable explanation.

The analysis of carbon and oxygen isotopes of 21 calcite samples from the fractures of 7 wells in the study area (Table 5-2) shows that their $\delta^{13}C$ test values (PDB standard) are between −9.18‰ and 0.18‰, averagely −4.70‰, and $\delta^{18}O$ test values (PDB standard) range between −15.77‰ and −9.96‰, averagely −13.13‰. Carbon and oxygen isotope composition of different calcite vein samples vary widely, which reflects the complexity of geological fluid in the Carboniferous-Permian fractures.

Table 5-2 Test data of carbon and oxygen isotopes of calcite samples in fractures in the study area and calculated fluid temperature.

No.	Sample No.	Filling	Horizon	Well depth /m	$\delta^{13}C_{PDB}$ /‰	$\delta^{18}O_{PDB}$ /‰	Fluid temperature $T/°C$
1	HQ101-3	Calcite	C	1739	−4.09	−14.53	100.68
3	HQ101-6	Calcite	C	1740.88	−5.00	−15.63	108.82
4	HQ101-7	Calcite	P	2237	−1.46	−10.07	70.18
7	HQ101-9-2	Calcite	P	2240.7	−2.38	−10.28	71.53
9	HQ102-2	Calcite +quartz	C	1338.11	4.07	−3.34	31.68
12	HQ3-2-1	Calcite	C	1075.92	1.06	−13.07	90.26
14	HQ3-3	Calcite +quartz	C	1478.19	−2.77	−12.83	88.59
16	HQ4-2	Calcite	C	686.4	−2.97	−14.46	100.18
19	HQ6-2	Calcite	C	255.5	4.42	−14.86	103.10
20	HQ6-3	Calcite	C	256	−3.73	−18.23	129.01
21	HQ6-4	Calcite	C	275.8	7.65	−19.13	136.32
22	HQ6-8	Calcite	P	1920.65	−3.83	−10.97	76.01
18	HQ6-13	Calcite +quartz	P	2702.14	0.17	−21.41	155.55
23	HQ7-1	Calcite	C	162.27	−3.66	−10.58	73.47
24	HQ7-2	Calcite	C	211.66	−7.05	−11.72	81.00
25	HQ7-3a	Calcite	C	212	−7.73	−16.76	117.43
26	HQ7-3b	Calcite	C	212	−11.04	−18.21	128.85
31	HS1-4	Calcite	P	2099.1	−1.18	−14.10	97.57
33	HS1-8	Calcite	P	2101.33	−2.14	−18.44	130.70
27	HS1-10	Calcite	P	2153.7	−4.30	−11.07	76.67
28	HS1-15C	Calcite	P	2553.5	3.54	−4.00	35.05

To figure out the sources of C and O isotopes in the carbonate rock filling the fractures in the study area, the carbon and oxygen isotopes source analysis chart proposed by Clark and Fritz (1997) and Hoefs (1997) were used to analyze the falling points of the isotope values of the test samples. From the falling point chart (Table 5-2) of $\delta^{13}C$ and $\delta^{18}O$ of the samples we can see that in the $\delta^{13}C$ falling point area, carbon in the fracture-filling calcite has multiple sources: possibly carbonate/diamond, atmospheric CO_2, mantle, marine carbonate, terrestrial carbonate and formation water. But in specific areas, other carbon sources (atmospheric

CO_2, mantle fluid, etc.) may also have significant influence. From the distribution of $\delta^{13}C$ value of the studied samples in the Hala'alate Mountain area (Figure 5-4), we can see that carbon sources of the Carboniferous-Permian formation fluid in the study area are various.

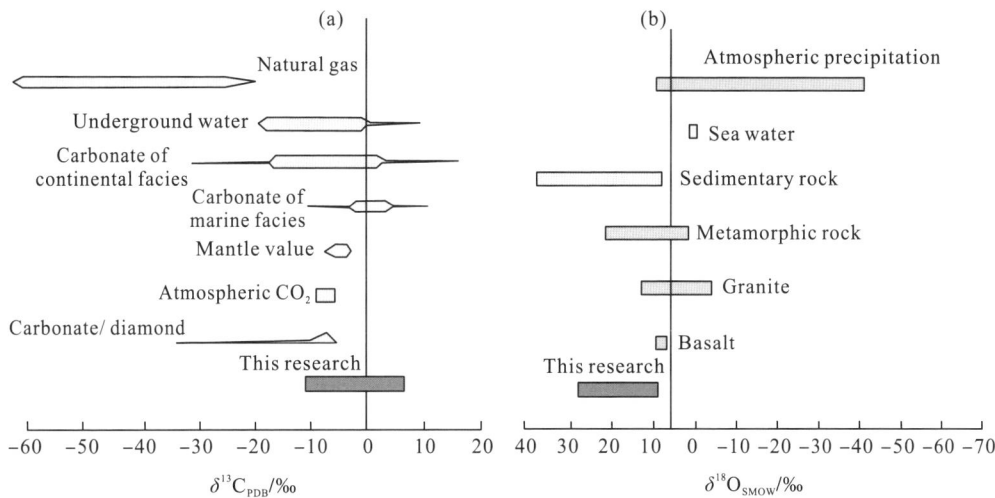

Figure 5-4　Important geological reservoirs: characteristics of carbon and oxygen isotopes (Hoefs, 1997; Clark and Fritz, 2000).

Compared with carbon isotopes, oxygen isotopes are more sensitive to temperature in isotope fractionation, and thus oxygen isotopes are often used by geologists as geological thermometer. Previous researches have proved that the high temperature of diagenesis and metamorphism will lead to large consumption of $\delta^{18}O$ in the water, and eventually negative $\delta^{18}O$ value of calcite filling reservoir fractures (Wang and Zhang, 2001; Jrgen, 1988). However, there is another reasonable explanation to this phenomenon: calcite formed in fresh water under low temperature. In general, the oxygen isotope composition characteristics of calcite filling fractures are controlled by fluid source and diagenetic temperature (dominant) (Shi et al., 2013).

From the $\delta^{18}O$-$\delta^{13}C$ (Figure 5-5) falling point chart of the Carboniferous-Permian calcite filling fractures we can see that the calcite in structural fractures in the study area is complex in genesis, possibly including low temperature alteration of volcanic rocks, contamination of sedimentary rock, decarboxylation of organic matter, and dissolution of carbonate rock, etc.

According to the previous $\delta^{13}C_{PDB}$ and $\delta^{18}O_{SMOW}$ value chart (Cao et al., 2007), and the falling points of the C and O isotope values in this test, we can find that the falling points of most samples from the study area fall in the mixing genesis section (Figure 5-6), which also suggests that the fluid that forms the vast majority of fracture-filling carbonate in the study area is not single, and the carbonate is often of mixed genesis or has suffered late transformation by fluids of different natures.

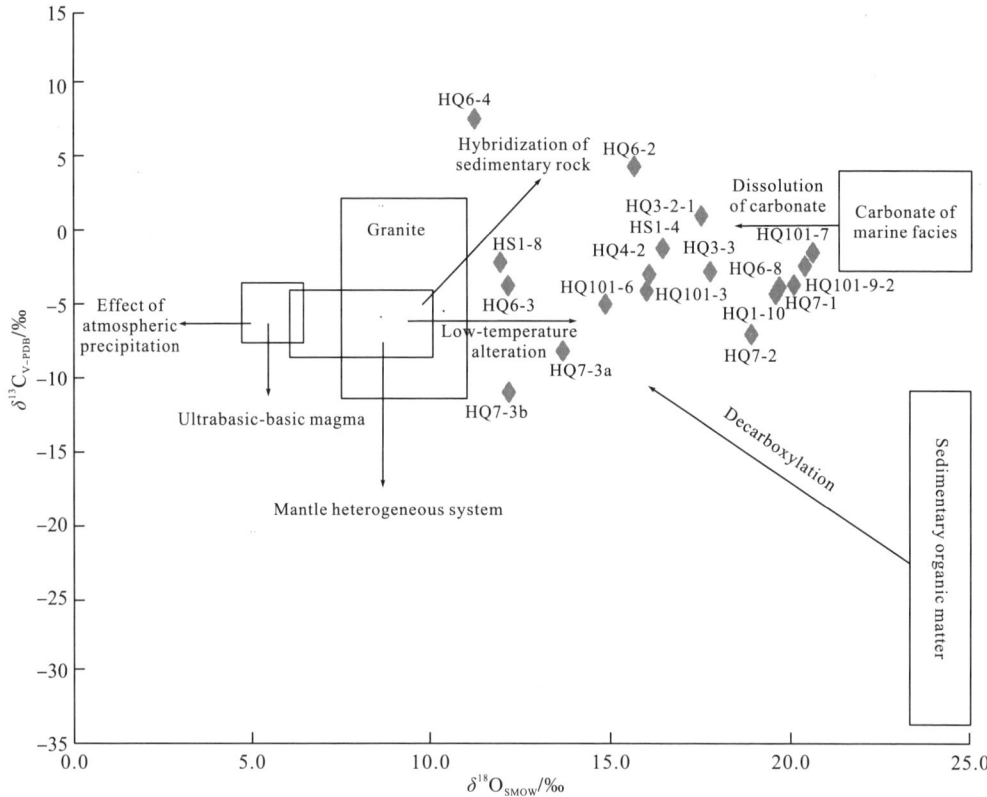

Figure 5-5　Geneses of $\delta^{18}O_{SMOW}$-$\delta^{13}C_{PDB}$ of the Carboniferous-Permian calcite in the Hala' alate Mountain area.

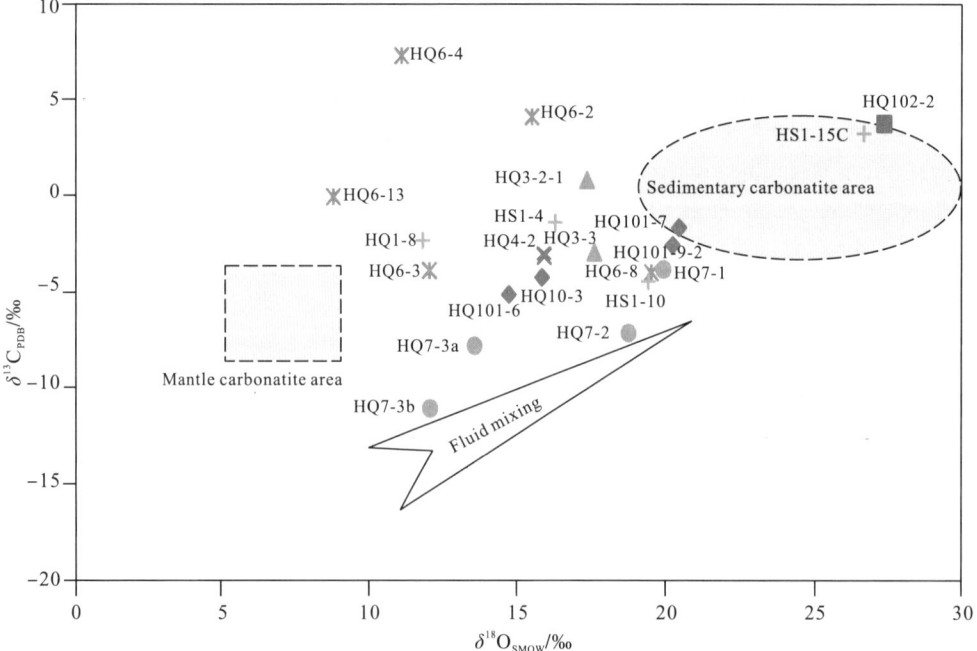

Figure 5-6　Relationship between $\delta^{13}C_{PDB}$ and $\delta^{18}O_{SMOW}$ values of calcite veins in the study area (boundary values of the chart from Hoefs, 1997; Toyoda, 1994; Taylor, 1976; Veizer, 1976).

It is difficult to infer the genesis of carbonate only based on the falling point of relevant $\delta^{13}C_{PDB}$ and $\delta^{18}O_{SMOW}$ value chart in Figure 5-4, Figure 5-5 or Figure 5-6, because there are many factors affecting C and O isotopes. Therefore, it is necessary to combine the falling point analysis with other analysis methods to get the most reliable interpretation on the genesis of carbonate.

Under the constraint of obtained interpretation results of strontium isotope analysis, the fluid source and genesis reflected by the carbon and oxygen isotope characteristics of Carboniferous and Permian fracture-filling calcite in the Hala'alate Mountain area have been analyzed further. According to the carbon and oxygen isotope data obtained from the test, the distribution characteristics of the calcite of three geneses in strontium isotope interpretation are obviously different on the falling point chart of C and O isotopes (Figure 5-7).

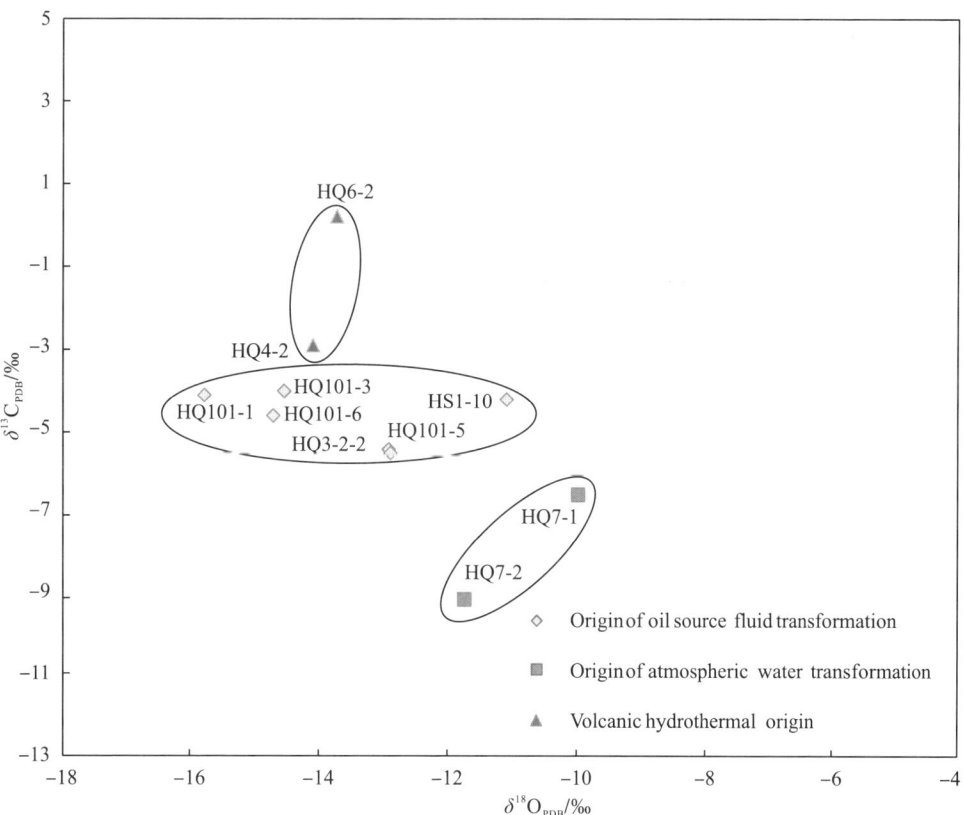

Figure 5-7 Relationship between $\delta^{13}C_{PDB}$ and $\delta^{18}O_{PDB}$ of calcite veins in the study area.

Sample HQ6-3 and HQ4-2 have a tested carbon isotope of 3.73‰ and −2.97‰ respectively, near the normal mantle source $\delta^{13}C$ value of −5‰ (Hoefs, 1997), and oxygen isotopic value of −14.46 ‰ and −18.23 ‰ respectively, the high negative values reflect that the fluid medium forming the calcite is high in temperature. The interpretation results of carbon and oxygen isotopes and strontium isotope are consistent, indicating the calcite formation is related to volcanic hydrothermal fluid.

Sample HQ7-1 and HQ7-2 has a carbon isotope of −7.05 ‰ and −6.62 ‰ respectively, near the $\delta^{13}C$ value of atmospheric CO_2 (on average −7‰) (Koch et al., 1992), and oxygen isotopic value of −11.72‰ and −9.95‰ respectively. Compared with volcanic hydrothermal calcite, the negative level of their oxygen isotopes decreases, the reason may be that the fractionation effect of temperature on oxygen isotope makes the calcite of shallow burial after superimposed transformation of atmospheric water under low temperature more enriched in ^{18}O. Therefore, after the transformation of atmospheric water, the volcanic hydrothermal calcite tends to get lighter in carbon isotope value and heavier in oxygen isotope value.

The tested sample HQ101-1, HQ101-3, HQ101-5, HQ101-6, HS1-10 and HQ3-2-2, have a carbon isotope between −4.09‰ and −5.52‰, and oxygen isotope between −12.90‰ and −15.77‰. Compared with volcanic hydrothermal genesis calcite, this kind of calcite originated from oil source fluid transformation tends to be more negative in carbon isotope, which may be caused by superimposed transformation of the late oil source fluid in the area to the early volcanic hydrothermal genesis calcite. Because the filling time of the latest oil source fluid is late (Cretaceous) (Wu, 2009), the calcite have low carbon isotopic exchange level, and the organic carbon mixed into calcite is insufficient to cause large negative offset on the carbon isotope of the samples. In addition, the test value of oxygen isotopes of the above samples are very close to each other and slightly heavier than the oxygen isotope of the volcanic hydrothermal genesis calcite samples overall, which may also result from oxygen isotope fractionation effect caused by the lower temperature of oil source fluid in later activity than that of volcanic hydrothermal fluid.

Since water medium temperature has much stronger effect on the $\delta^{18}O$ value than salinity, while the $\delta^{13}C$ value changes little with temperature. Therefore, when the salinity is constant, the $\delta^{18}O$ value can be used as a reliable marker of ancient temperature. When the carbonate and water medium are in a state of balance, the $\delta^{18}O$ value will drop with the rise of temperature (Zhang, 1985). The method of measuring water temperature of ancient ocean with $\delta^{18}O$ was proposed by Urey (1934), an American scholar and Nobel Prize winner, and materialized by Epstein and Ofhers (1951, 1953). Shackleton (1974) further modified it and obtained the final empirical formula:

$$T(°C) = 16.9 - 4.38 \times (\delta C - \delta W) + 0.10 \times (\delta C - \delta W^2)$$
$$\delta C = 10.25 + 1.01025 \times \delta CaCO_3, \quad \delta W = 41.2(\delta H_2O = 0)$$

The mineralization temperature of the collected samples were calculated with the above carbonate mineralization temperature calculation formula and the carbon and oxygen isotope test results (Table 5-2). According to the calculated mineralization temperature and the statistical formation temperature range of the fracture-filling carbonate minerals (Table 5-2 and

Figure 5-7), the formation temperatures of carbonate minerals filling Carboniferous-Permian structural fractures in the study area are mainly distributed in three intervals (i.e., 30-70℃, 70-80℃ and 100-160℃), which suggests that fluids forming the fracture-filling carbonate have at least three temperature ranges, and that the fracture-filling carbonate has multiple fluid geneses. 30-70℃ represents the mineral-forming fluid related to the near-surface atmospheric fresh water or lake water; 70-80℃ represents the mineral-forming fluid discharged by the hydrocarbon source rock that was about to enter hydrocarbon generation threshold; 100-160℃ represents the fluid discharged by the hydrocarbon source rock reaching the hydrocarbon generation peak or fluid related to the deep mantle source thermal fluid.

In addition, based on the above strontium isotope analysis and observation of actual drilling cores and inclusion slices, it is found that core samples with fluid temperature calculated by the C and O isotopes of 70-80℃ and 100-160℃ tend to have good oil and gas shows, in particular, the samples taken from Well HQ101. This implies that the oil source fluid moving in the fractures of the study area may be of multiple phases.

However, since the fracture-filling carbonate in the study area has a variety of geneses, and different proportions of carbonate of one genesis or transformation will affect the $\delta^{13}C$ and $\delta^{18}O$ isotope values of test samples, it is clearly irrational to think that the $\delta^{18}O$ value is controlled by formation temperature alone. Therefore, the formation temperature of carbonate minerals filling fractures calculated by C and O isotopes probably cannot represent the real formation temperature of the minerals. The oil source fluid temperature associated with hydrocarbon accumulation in our interest should be found in combination with inclusion temperature measurement further.

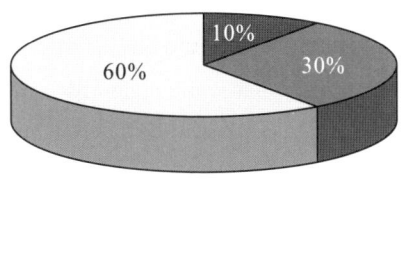

Figure 5-8 Forming temperature range of Carboniferous-Permian calcite veins.

5.2 Fluid inclusion geochemistry

Since Sorby et al. found inclusions in a variety of forms in quartz, topaz and other minerals in the mid-19th century, researchers have got a better understanding on and comprehensive and scientific definition of fluid inclusion in the course of detailed research. Fluid inclusions are the part of diagenetic mineralization fluids (gas and liquid fluid or silicate molten mass) wrapped in mineral crystal lattice defects or caves in the process of mineral crystal growth and now still sealed in the host minerals with a phase boundary with the host minerals. Fluid inclusions contain various original geochemical information of geological environment during the mineral formation (e.g., P, T, PH, X, salinity, etc.), and can be divided into primary inclusions, secondary inclusions and pseudosecondary inclusions according to the genesis; organic inclusions and mineral inclusions according to the composition; and pure gas inclusions, gas inclusions and fluid inclusions according to the gas-liquid ratio (Liu 2008).

Carboniferous-Permian structural fractures widespread in the Hala'alate Mountain area provide good migration pathways for fluid activities in the area, and fluids would leave marks in the fractures (i.e., fillings) in the course of activity. Therefore, studying the nature, source and capture time of the ancient fluids captured in fracture fillings can give us a clearer understanding of the activity characteristics of ancient fluids. Because there are multiphase hydrocarbon charging in the northwestern margin of the Junggar Basin, the analysis on fluid inclusions in the fracture fillings there is of great significance for the analysis of Carboniferous-Permian hydrocarbon accumulation period in the Hala'alate Mountain area.

5.2.1 Inclusion petrography

Fluid inclusions are widespread in the Carboniferous-Permian reservoirs in the study area. Based on the relationship between the formation time of the fluid inclusions and fracture fillings as well as the distribution characteristics of fluid inclusions under microscope, fluid inclusions in the area can be divided into inclusions of primary genesis and inclusions of secondary genesis. Observation of reservoir fluid inclusions in the study area (Table 5-3) reveals that inclusions of the same phase in different samples have basically similar development characteristics, while inclusions of the different phases in the same sample vary greatly in the development phase state and GOI. The first-phase primary inclusions have low GOI value, generally in the range of 5%–6%, and hydrocarbon fluid inclusions are mainly composed of liquid hydrocarbon inclusions, with a development ratio of above 80%; the second-phase secondary inclusions have higher GOI value (up to 80%), and the hydrocarbon fluid inclusions can be divided into liquid hydrocarbon inclusions, gas hydrocarbon inclusions and gas and liquid hydrocarbon inclusions according to the fluid phase state. This suggests that crude oil captured by the second-phase secondary inclusions is much higher in maturity than crude oil captured by the first-phase primary inclusions in the reservoirs of the Hala'alate Mountain area.

Table 5-3 Characteristics of fluid inclusions in the reservoirs of the Hala'alate Mountain area.

Well No.	Horizon	Depth/m	Period of time	Abundance	GLR	GOI	Phase state and proportion
HQ102-2	C	1338.1	Phase 1	High	≤5%	20%	Liquid hydrocarbon:80%; gas hydrocarbon:15%; gas-liquid hydrocarbon:5%
			Phase 2	Low	≤5%	1%-2%	Liquid hydrocarbon:70%; gas hydrocarbon:30%
HQ3-2-1	C	1075.92	Phase 1	High	≤5%	10%	Liquid hydrocarbon:90%; gas hydrocarbon:10%
HQ3-2-2	C	1075.92	Phase 1	High	≤5%	5%-6%	Liquid hydrocarbon:80%; gas hydrocarbon:20%
			Phase 2	High	≤5%	1%-2%	Liquid hydrocarbon:90%; gas hydrocarbon:10%
HQ3-3	C	1478.19	Phase 1	High	≤5%	5%-6%	Liquid hydrocarbon:90%; gas hydrocarbon:10%
HQ6-3	C	256	Phase 1	High	≤5%	5%-6%	Liquid hydrocarbon:80%; gas hydrocarbon:20%
HS1-4	C	2099.1	Phase 2	High	≤5%	80%	Liquid hydrocarbon:40%; gas-liquid hydrocarbon:60%
HS1-8	C	2101.33	Phase 2	High	≤5%	80%	Liquid hydrocarbon:20%; gas-liquid hydrocarbon:80%

Distributed in groups or sparsely in the fracture calcite veins and quartz veins, the primary inclusions can be divided into five categories: brown-dark brown liquid hydrocarbon inclusion, yellowish-gray gas-liquid hydrocarbon inclusion, dark gray gas hydrocarbon inclusion, yellowish-gray hydrocarbon-bearing brine inclusion and transparent colorless single-phase brine inclusion. The primary organic inclusions have high abundance, and GOI of 5%-6%, in which liquid hydrocarbon inclusions account for 80% and gas hydrocarbon inclusions account for 20%. Primary inclusions are regular in shape and various in size, mainly 3-15 μm and occasionally greater than 15 μm. The hydrocarbon-bearing brine inclusions associated with primary organic inclusions are no more than 5% in gas liquid ratio (Figure 5-9).

Secondary inclusions account for a large proportion in the fluid inclusions of the Carboniferous-Permian reservoir in the study area. The secondary inclusions are in zonal distribution along calcite-quartz microcracks, in which abundant brown and dark brown liquid hydrocarbon inclusions and yellowish-gray gas-liquid hydrocarbon inclusions are visible. Secondary organic inclusions have high abundance, and a GOI of 80%, in which liquid hydrocarbon inclusions account for 20% and gas-fluid hydrocarbon inclusions account for 80%. The secondary inclusions have regular shape and bigger size than the primary inclusions, mainly 5-12 μm and occasionally greater than 20 μm. The secondary gas-liquid hydrocarbon inclusions are 5% or less in gas liquid ratio (Figure 5-10).

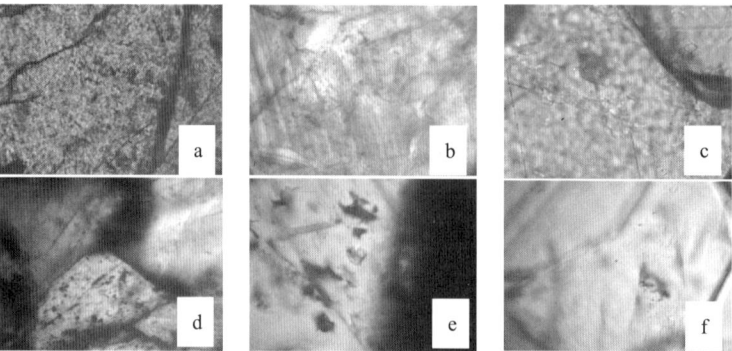

Figure 5-9 Microscopic characteristics of primary hydrocarbon fluid inclusions. a: Liquid hydrocarbon inclusions distributed in groups in calcite, Carboniferous, volcanic breccia, Well HQ3, 1475.92 m; b and c: Gas-liquid hydrocarbon inclusions scattered in calcite, Carboniferous, volcanic breccia, Well HQ102, 1338.11 m; d: Hydrocarbon-bearing brine inclusions, liquid hydrocarbon inclusions and gas hydrocarbon inclusions distributed in groups in calcite, Carboniferous, volcanic breccia, Well HQ3, 1478.19 m; e and f: Hydrocarbon-bearing brine inclusions and gas hydrocarbon inclusions distributed in groups, Carboniferous, tuff, Well HQ6, 256 m.

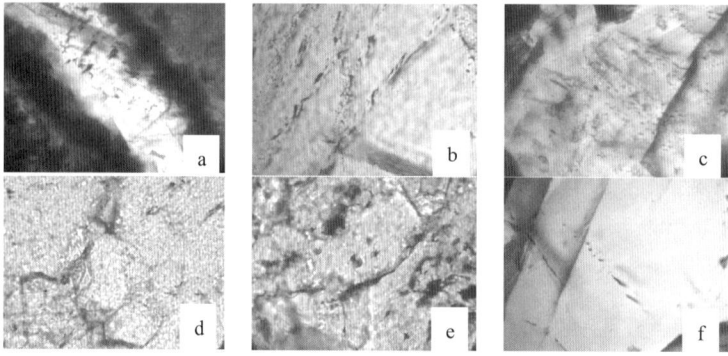

Figure 5-10 Microscopic characteristics of secondary hydrocarbon fluid inclusions. a, b and c: Liquid hydrocarbon inclusions and gas hydrocarbon inclusions in zonal distribution in calcite, Carboniferous, volcanic breccia, Well HQ3, 1475.92 m; d: Liquid hydrocarbon and gas-liquid hydrocarbon inclusions in zonal distribution in calcite, Permian, mudstone, Well HS1, 2099.1 m; e: Gas-liquid hydrocarbon inclusions in zonal distribution in calcite, Permian, mudstone, Well HS1, 2101.33 m; f: Dark gray gas hydrocarbon inclusions in zonal distribution in quartz minerals, Carboniferous, tuff, Well HQ6, 256 m.

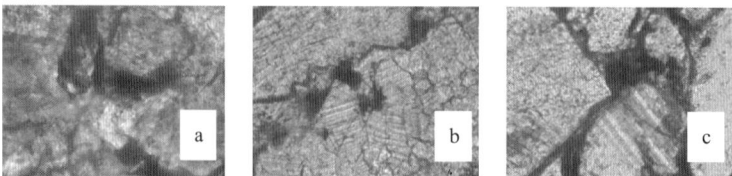

Figure 5-11 Hydrocarbon filling characteristics of Carboniferous fracture calcite vein and quartz vein microcracks. a: Asphalt of brown thin oil filling the mcrocracks between calcite crystals, Carboniferous, volcanic breccia, Well HQ102, 1338.11 m; b and c: Brown asphalt of thin oil filling the mcrocracks between calcite crystals, Carboniferous, volcanic breccia, Well HQ3, 1475.92 m.

Besides primary and secondary hydrocarbon fluid inclusions, there are also asphalt fillings in clacite and quartz veins in Carboniferous-Permian rock in the study area (Figure 5-11). This evidence suggests that Carboniferous-Permian reservoir in the area experienced multiphase hydrocarbon filling process.

5.2.2 Fluorescence characteristics

Organic inclusions under ultraviolet light, purple light or blue light emit lights with longer wavelength than that of irradiation beam in a very short time, namely the fluorescence of organic inclusions. When organic molecules were irradiated by the high-energy shorter-wavelength light, the light quantum would make the electron in the ground state excited and jump to the track of higher energy level. Electrons on the track of high energy level are not stable, and go back to the ground state through the release of energy by emitting corresponding light quantum. As a result, fluorescence is generated. Fluorescence generation of organic inclusions requires certain condition, i.e., the organic molecules contain conjugated double bond, the greater the conjugate degree, the easier it will be to be stimulated to produce fluorescence. Therefore, the vast majority of inclusions that can emit fluorescence usually contain aromatic ring or heterocyclic compound of conjugated double bond molecules (Gao, 2000), so brine inclusions and gas inclusions generally do not emit fluorescence.

Multiphase hydrocarbon fluid inclusions in the fracture-filling calcite of the Carboniferous-Permian reservoir in the study area are direct evidence of hydrocarbon charging in the area, and their fluorescence characteristics reflect the maturity of oil and gas contained in the organic inclusions. With the increase of maturity, fluorescence excited by blue light in hydrocarbon fluid inclusions changes from dark brown to orange, yellow green and then dark green successively (Gao et al., 2015) Primary liquid hydrocarbon inclusions in the Carboniferous-Permian reservoir in the study area give out light yellow green fluorescence under the excitation of blue light, while primary gas hydrocarbon inclusions show no fluorescence (Figure 5-12).

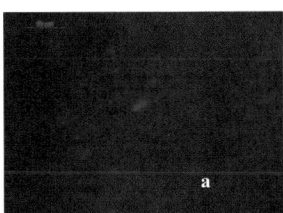
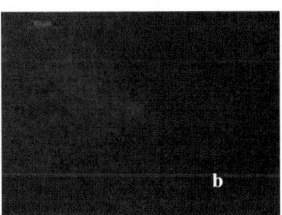

Figure 5-12 Fluorescence characteristics of primary hydrocarbon fluid inclusions in the Carboniferous reservoir. a and b: Gas-liquid hydrocarbon inclusions distributed sporadically in calcite, light yellow green fluorescence show, Carboniferous, volcanic breccia, Well HQ102, 1338.11 m.

Observation of the fluorescence of hydrocarbon fluid inclusions in Carboniferous-Permian reservoir under microscope shows that secondary liquid hydrocarbon inclusions and gas-liquid hydrocarbon inclusions in the reservoir give out light yellow green fluorescence under the excitation of blue light (Figure 5-13), and their fluorescent display intensity and number are significantly higher than those of primary hydrocarbon fluid inclusions. In addition, the part of the fracture of calcite veins and quartz veins impregnated by crude oil give out light aquamarine blue fluorescence or no fluorescence (Figure 5-14).

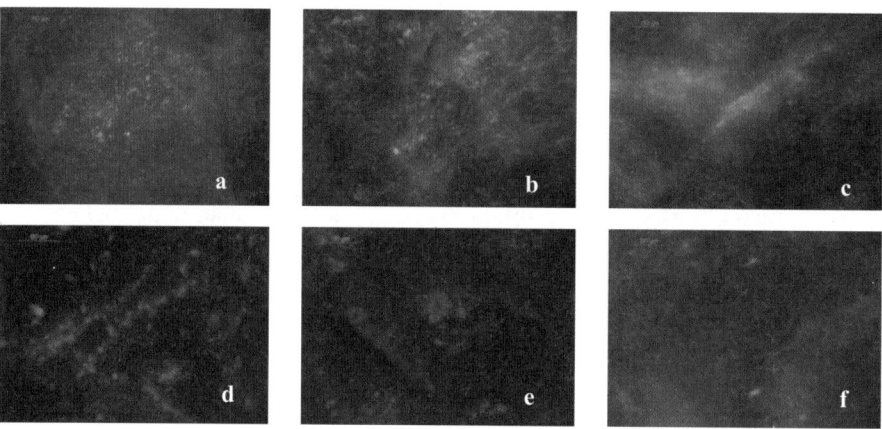

Figure 5-13 Fluorescence characteristics of secondary hydrocarbon fluid inclusions. a, b and c: Liquid inclusions and gas liquid hydrocarbon inclusions in zonal distribution in the microcracks of calcite veins and quartz veins, showing pale green and pale yellow green fluorescence, Permian, mudstone, Well HS1, 2099.1 m; d, e and f: Liquid inclusions and gas liquid hydrocarbon inclusions in zonal distribution in the microcracks of quartz veins, showing pale yellow green fluorescence, Permian, mudstone, Well HS1, 2101.33 m.

The fluorescent display characteristics of above hydrocarbon fluid inclusions indicate that the oil and gas filling Carboniferous-Permian reservoir are in medium-high maturity stage.

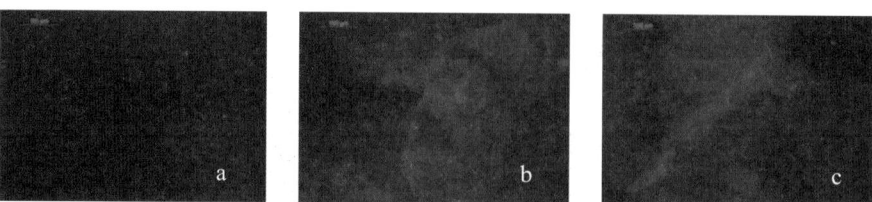

Figure 5-14 Fluorescence characteristics of hydrocarbon filling Carboniferous fracture calcite vein and quartz vein microcracks. a: Brown thin oil asphalt filling the mcrocracks between calcite crystals, no fluorescence display, Carboniferous, volcanic breccia, Well HQ102, 1338.11 m; b: Brown thin oil asphalt filling the mcrocracks between calcite crystals, no fluorescence display, Carboniferous, volcanic breccia, Well HQ3, 1475.92 m; c: The mcrocracks between calcite crystals with crude oil impregnation showing pale yellow and pale yellow green fluorescence, Carboniferous, volcanic breccia, Well HQ3, 1478.19 m.

5.2.3 Homogenization temperature

Fluid inclusions in hydrocarbon-bearing basins are mainly divided into two types: one is the fluid inclusion captured by the hydrocarbon source rock and reservoir from a homogeneous fluid system during the diagenetic evolution; the other is the mixed phase inclusion captured from a variety of non-homogeneous fluid systems (Gao and Chen, 2000). The former is the main object of traditional inclusion homogenization temperature measurement. "Homogenization temperature" refers to the instantaneous temperature when the inclusion in two phases or multiphase at room temperature is transformed into the original uniform single-phase fluid through artificial heating. Homogenization temperature is generally regarded as the formation temperature at the time when the mineral captures the inclusion in the homogenous fluid system. By combining homogenization temperature measurement with reconstruction of stratigraphic burial-thermal evolution history, it is able to reconstruct the information about the fluid activity period and the hydrocarbon accumulation process in a region (Zhang, 2010).

The existing research results of the northwestern margin of the Junggar Basin show that the activity of hydrocarbon fluids are closely related to important tectonic events in the area. According to previous research results of the reconstructed burial history and hydrocarbon accumulation periods, there are three major hydrocarbon filling periods of hydrocarbon accumulation in the northwestern margin, which are corresponding to three important stages of tectonic evolution: thrust nappe development period (late Permian-early Triassic), compression-torsion deformation transformation period (late Triassic-early Jurassic), and weak transformation period of basin depression (late Jurassic-early Cretaceous).

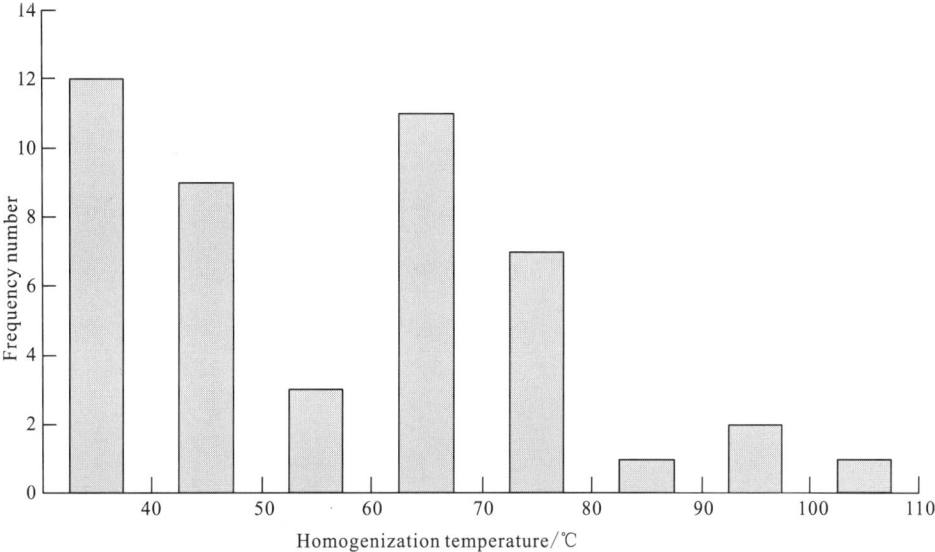

Figure 5-15 Distribution of homogenization temperature of hydrocarbon-bearing brine inclusions in Carboniferous-Permian reservoir of the Hala'alate Mountain area.

The above oil-source correlation analysis for the Hala'alate Mountain area shows that the Carboniferous-Permian oil and gas in the area mainly came from the Permian hydrocarbon source rock. According to the statistics of the homogenization temperature of hydrocarbon-bearing brine inclusions in Carboniferous-Permian reservoir, the homogenization temperatures are mainly distributed in three ranges: 36-52℃, 56-79℃ and 87-108℃ (Figure 5-15). The research suggests the accumulation fluid activity in Carboniferous-Permian reservoir of the Hala'alate Mountain area is comparable with that in other areas of the northwestern margin of the Junggar Basin, and also went through the above three periods of hydrocarbon filling and accumulation (late Permian-early Triassic, late Triassic-early Jurassic, and late Jurassic-early Cretaceous).

Chapter 6
Mechanical Properties of Reservoir Rocks

Tectonic deformation research, calculation of tectonic stress field, and development and forecast of reservoir fractures are all closely related to the mechanical properties of rock. Mechanical properties of rock refer to the deformation characteristics of rock under the act of stress. Parameters commonly used to indicate mechanical properties of rock include compressive strength, Young's modulus, and Poisson's ratio, etc. A lot of experimental results and theoretical studies have proven that a host of factors influence the mechanical properties of rock, such as rock type, confining pressure, temperature, pore pressure and medium, strain rate, stress state, etc. The formation conditions affecting rock mechanical property most include formation temperature, confining pressure, fluid saturation, direction of the bedded rock, etc. The formation of fissure is related to the internal factors such as the lithology of the reservoir in addition to the palaeotectonic stress field. These internal factors affect the density and distribution, etc. of the fissures. Fissure development is dependent on both external cause (tectonic deformation) and internal cause, hereinto, lithology is the internal cause that controls the fissure development potential (becoming reservoir). Lithology includes two aspects: rock composition and rock grain size. In general, under the same conditions, the degree of fissure development will be higher in the rock with higher brittle components (such as quartz, calcite, feldspar and other brittle minerals) (Ju W et al., 2014; Cai and Tong, 2010).

These parameters can be obtained only in the laboratory test, which then will be compared with the actual geological conditions and explained. Therefore, the rock mechanics parameters obtained from conventional test conditions are different from that in the real geological conditions, and such difference is often regional, and thus empirical formula has no universal significance.

The Carboniferous and Permian in the Hala'alate Mountain area are all relatively dense rock formations and thus the development degree of the fissures is particularly important for oil and gas migration and accumulation. Both in forecasting reservoir fracture and analyzing the tectonic deformation style, the research of mechanical properties of rock is of great importance for the oil and gas exploration in this area.

6.1 Rock mechanics characteristics

6.1.1 Sample information and test

In the research, mechanics tests of reservoirrock in the Hala'alate Mountain area were all conducted in the State Key Laboratory of Oil and Gas Reservoir Geology and Exploitation in Chengdu University of Technology. The rock mechanics feature experiments were done on "MST Rock Physical Parameter Testing System" developed by the United States MST (Figure 6-1). The system consists of three subsystems: digital electro-hydraulic servo subsystem for rigid rock mechanics test, rock ultrasonic measurement subsystem, and rock pore volume change and permeability test subsystem, which can test rock mechanics parameters, physical parameters and ultrasonic velocity under simulated formation conditions (temperature: room temperature to 200 ℃, confining pressure: 0-140 MPa, pore pressure: 0-70 MPa, axial force: 0-1600 kN). In this study, the rock mechanics test subsystem of this system was used to test rock mechanics parameters of rock samples from Carboniferous-Permian reservoirs in the Hala'alate Mountain area. The experiment content includes uniaxial tensile strength, uniaxial compressive strength, triaxial compressive strength, deformation modulus, and Poisson's ratio, etc.

Figure 6-1 MST Rock Physical Parameter Testing System.

The samples for rock mechanics characteristics test were taken from 5 wells (HS1, HS2, HQ3, HQ6, and HQ101) in the study area, including mudstone, tuff, volcanic breccia and basalt, and covering the Carboniferous and Permian (Table 6-1). The test samples can basically reflect the rock types and rock mechanics characteristics of the Carboniferous-Permian reservoir in the study area.

Table 6-1 Basic information of Carboniferous–Permian samples for rock mechanics testing in the Hala'alate Mountain Mountains.

Sampled well	Depth/m	Horizon	Rock type	Test type	
				Tensile	Compressive
HS1	2097.4	P	Mudstone	√	√
HS1	2151.2	P	Mudstone	√	√
HS1	2153.5	P	Mudstone	√	√
HS1	2155.6	P	Mudstone	√	√
HS1	2102.63	P	Mudstone	—	√
HS2	1210.2	C	Tuff	√	—
HQ3	2796	C	Tuff	√	√
HQ3	1210.3	C	Volcanic breccia	—	√
HQ3	1478.19	C	Volcanic breccia	—	√
HQ3	2687.1	C	Tuff	—	√
HQ6	275.8	C	Tuff	—	√
HQ6	2545.5	P	Mudstone	√	—
HQ6	2700.93	P	Mudstone	√	—
HQ101	1436.11	C	Fine conglomerate	√	√
HQ101	2236.15	P	Mudstone	√	√
HQ101	2241	P	Mudstone	—	√
HQ101	2242.21	P	Mudstone	—	√

6.1.2 Sample test procedure and results

To meet the experimental requirements of "MST Rock Physical Parameter Testing System", the test samples were all drilled into cylinder specimens 25 mm in diameter and 50 mm high, with the end faces polished. After processed and air dried, these samples were sealed in heat shrinkable sleeve and placed in high temperature and high pressure triaxial testing apparatus, were installed with high precision extensometer that measured the longitudinal and lateral deformation. Firstly, the rock samples were applied with the confining pressure (hydrostatic pressure) and pore pressure to the set value, and heated to set temperature 25 ℃. After the confining pressure and temperature were stable, they were applied with axial stress (differential stress) at equal axial displacement rate until the samples were damaged. According to the records of axial stress, axial deformation and lateral deformation data, we could obtain stress-strain characteristics in the process of rock compression and rock mechanics parameters in the corresponding temperature pressure conditions. Table 6-2 shows the results of tensile strength of Carboniferous-Permian reservoir rock in the Hala' alate Mountain area in the uniaxial test and Table 6-3 shows the results of rock mechanics parameters in the triaxial test.

Table 6-2 Uniaxial test results of tensile strength of Carboniferous-Permian reservoir rock in the Hala' alate Mountain area.

Well No.	Horizon	Lithology	Average diameter/mm	Average height /mm	Mass /g	Failure load /kN	Tensile strength/MPa	Remarks
HS1	P	Mudstone	25.40	26.27	35.08	8.85	8.45	
HS1	P	Mudstone	25.37	26.21	35.2	9.93	9.51	
HS1	P	Mudstone	25.10	25.47	32.11	4.48	4.46	End defect
HS1	P	Mudstone	25.34	26.34	35.21	8.48	8.09	
HQ101	C	Volcanic breccia	25.37	25.20	31.96	6.71	6.68	
HQ101	P	Mudstone	25.35	25.44	32.5	5.677	5.61	
HQ3	C	Tuff	25.34	26.13	34.2	13.75	13.23	
HQ6	P	Mudstone	25.29	24.37	30.8	4.99	5.16	End defect
HQ6	P	Mudstone	25.39	26.31	35.51	5.19	4.95	
HS2	C	Tuff	25.36	25.70	33.36	9.22	9.01	

Chapter 6 Mechanical Properties of Reservoir Rocks

Table 6-3 Triaxial test results of rock mechanics parameters of Carboniferous-Permian reservoir.

Test sample No.	Sampled well	Lithology	Depth /m	Horizon	Confining pressure /MPa	Pore pressure /MPa	Temperature /°C	Saturation condition	Compressive strength /MPa	Modulus of deformation /GPa	Poisson's ratio	Remarks
3	HS1	Mudstone	2097.4	P	26.85	0	25	Air dry	475.13	55.9	0.266	
6	HS1	Mudstone	2102.63	P	26.91	0	25	Air dry	309.72	45.4	0.248	
7	HS1	Mudstone	2151.2	P	27.54	0	25	Air dry	327.74	54.3	0.274	
1	HS1	Mudstone	2153.5	P	27.56	0	25	Air dry	—	—	—	Damaged when applied with confining pressure!
16	HS1	Mudstone	2155.6	C	27.59	0	25	Air dry	241.88	55.2	0.265	
4	HQ101	Volcanic breccia	1436.11	C	18.95	0	25	Air dry	153.73	29.4	0.16	
8	HQ101	Mudstone	2236.15	P	29.95	0	25	Air dry	257.19	35	0.343	
5	HQ101	Mudstone	2241	P	30.03	0	25	Air dry	51.96	18.8	—	Fissure appears!
11	HQ101	Volcanic breccia	2242.21	C	30.05	0	25	Air dry	65.13	46.5	0.178	
13	HQ3	Volcanic breccia	1210.3	C	15.61	0	25	Air dry	164.13	35.8	0.16	
9	HQ3	Tuff	1478.19	C	19.36	0	25	Air dry	174.76	23.6	0.09	
12	HQ3	Tuff	2687.1	C	38.16	0	25	Air dry	256.36	43	0.186	
14	HQ3	Tuff	2796	C	39.7	0	25	Air dry	271.35	44.1	0.261	
2	HQ6	Tuff	275.8	C	3.34	0	25	Air dry	59.68	42.6	0.159	

6.1.3 Rock deformation characteristics

Static deformation refers to the process during which, the sample is applied with axial differential stress in a certain temperature condition to make the sample deform until it is damaged. During this process, the curves drawn for recording axial and radial strain to stress at all levels are called the statics deformation curves. Under normal circumstances, the whole process from the start of stress loading to the rock failure can be divided into elastic deformation stage, inelastic deformation stage (expansion stage) and (after) breaking stage. Analysis of the rock mechanics experiment results shows different Carboniferous-Permian rock types in the Hala'alate Mountain area in the process of the experiment of "MST Rock Physical Parameter Testing System" mainly had elastic-plastic deformation. In the following, we will plot the stress-strain curves according to the experimental results and analyze typical rock deformation characteristics of the main rock types in the study area in detail.

1. Mudstone

1) No.1 mudstone

No.16 sample is the Permian mudstone at the buried depth of 2155.6 m in Well HS1. Figure 6-2 shows the stress-strain curves of this sample under the confining pressure of 27.59 MPa. Its ultimate strength is 240 MPa. When the axial differential stress was less than 240 MPa, the curve was almost linear, which reflects the process of the elastic deformation of the rock sample. The yield stress is close to the rock ultimate strength and its plastic deformation before failure was very small, with a total deformation of less than 0.6%, representing typical brittle deformation characteristics.

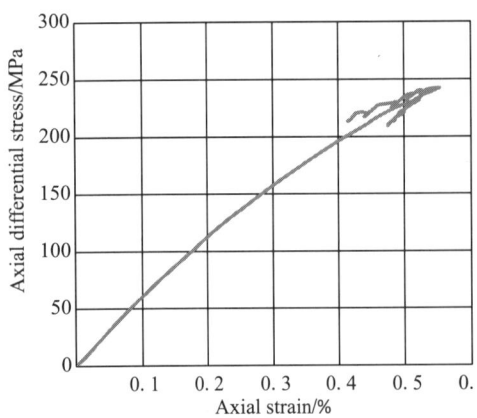 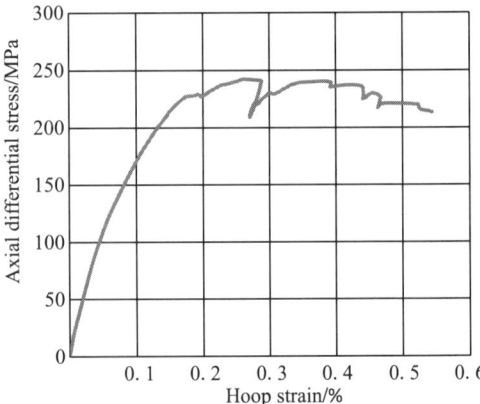

Figure 6-2　Stress-strain relationship of No.16 sample(Confining pressure: 27.59 MPa, HS1, Permian, mudstone, 2155.6 m).

2) No.3 mudstone

No.3 sample is Permian mudstone at the buried deep of 2097.4 m in Well HS1. Figure 6-3 shows the stress-strain curves of this sample under the confining pressure of 26.85 MPa. Its rock ultimate strength is 475 MPa. From the start of axial stress application, the stress-strain curve was nearly linear. When the axial differential stress reached 475 MPa, small plastic deformation appeared in the rock, the stress-strain curve slightly bended downward, and the curve slope was reduced until the rock was fractured. The strain deformation of this sample before failure was more than 1% and total lateral deformation exceeded more than 0.4%. Its stress-strain curve reflects the elastic-plastic deformation characteristics.

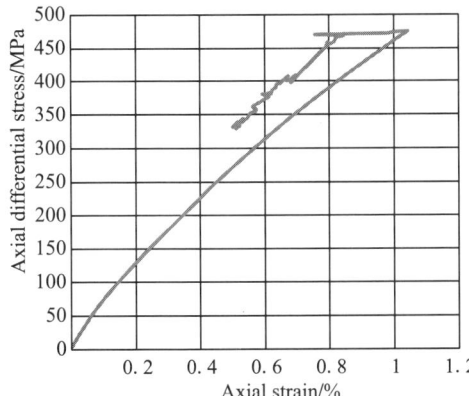 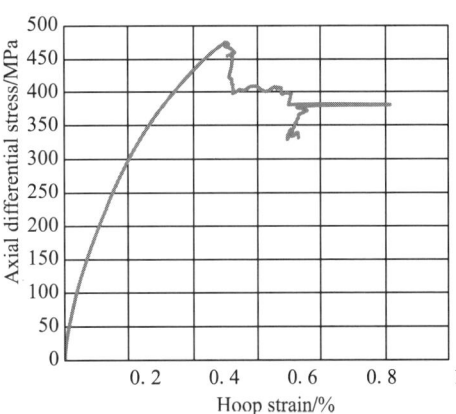

Figure 6-3   Stress-strain relationship of No.3 sample (Confining pressure: 26.85 MPa, HS1, Permian, mudstone, 2097.4 m).

3) No.8 mudstone

No.8 sample is Permian mudstone at the buried deep of 2236.15 m in HQ101. Figure 6-4 shows that the rock stress-strain curve consist of approximately linear elastic deformation stage and plastic deformation stage before rock failure. The yield point of this sample cannot be easily determined on the stress-strain curve. The sample has a larger compressive strength of 257.15 MPa. This sample had larger deformation in the process of test, with the axial strain of more than 1% and hoop strain close to 1%, showing elasto-plastic deformation characteristics.

4) No.11 mudstone

No.11 sample is the Permian mudstone at the buried depth of 2242.41 m in Well HQ101. Figure 6-5 stress-strain curve show that the ultimate strength of No.11 sample is 65.13 MPa under the confining pressure of 30.05 MPa. When the axial differential stress is less than 65.13 MPa, the axial strain and axial differential stress of the sample nearly increase linearly, reflecting elastic deformation process. When the axial differential stress is larger than 65.13 MPa, the stress-strain curve bends downward, reflecting the plastic deformation process of the sample. No.11 sample showed the elastic-plastic deformation characteristics in the process of rock mechanics test. Its deformation was larger, with axial deformation of more than 1%, and hoop deformation of more than 0.7%.

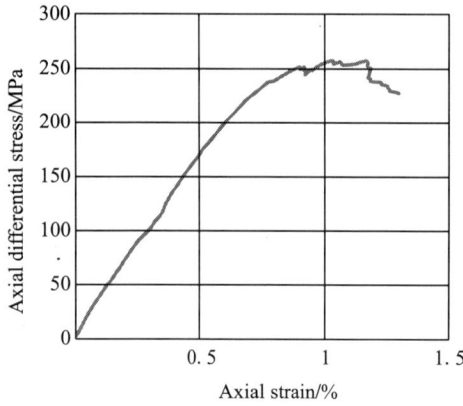 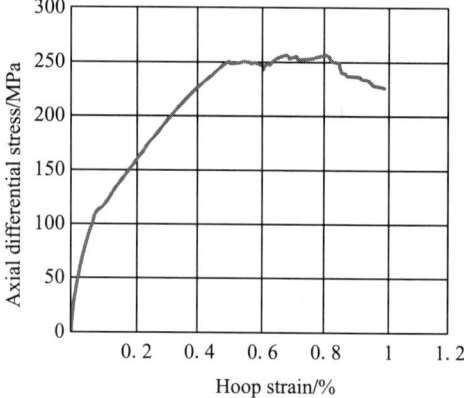

Figure 6-4 Stress-strain relationship of No.8 sample (Confining pressure: 29.96 MPa, HQ101, Permian, mudstone, 2236.15 m).

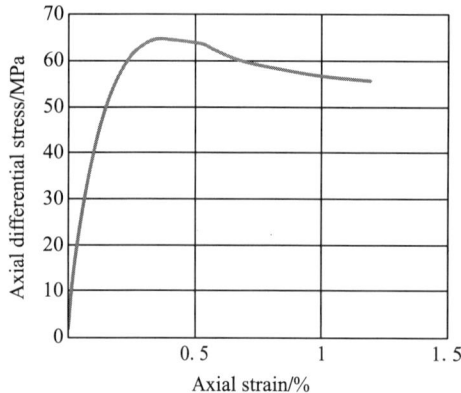 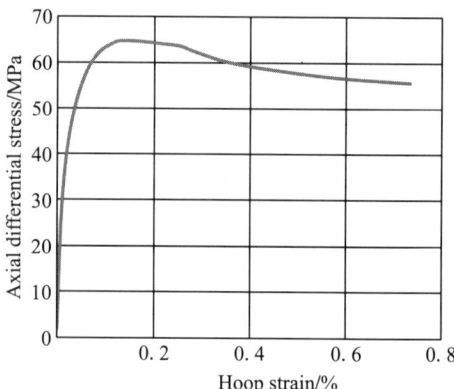

Figure 6-5 Stress-strain relationship of No.11 sample (Confining pressure: 30.05 MPa, HQ101, Permian, mudstone, 2242.21 m).

2.Tuff

1) No.2 tuff

No.2 sample is the Carboniferous tuff at the buried depth of 275.8 m in Well HQ6. Figure 6-6 stress-strain curve show that the rock experienced elastic deformation stage and plastic deformation stage before rock failure. When axial differential stress was less than 59.68 MPa, the sample showed elastic deformation. When reaching rock ultimate strength of 59.68 MPa, the rock fractured after slight plastic deformation. Overall deformation of the No.2 sample was small in the process of rock mechanics test, with an axial strain of less than 0.25% and hoop strain of 0.5%. Compared with the ultimate strength of the tuff of HQ3, this rock sample is significantly smaller in strength.

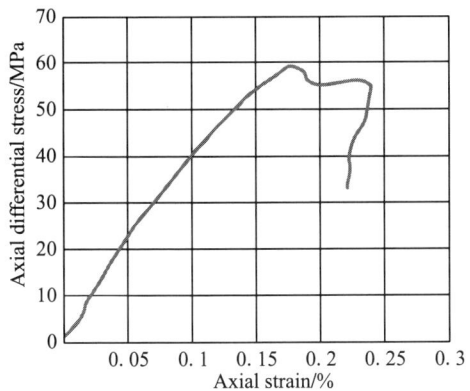
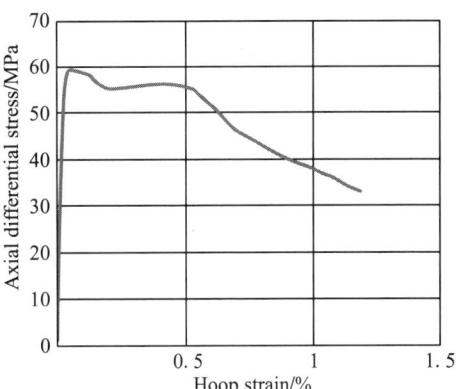

Figure 6-6 Stress-strain relationship of No.2 sample (Confining pressure: 3.34 MPa, HQ6, Carboniferous, tuff, 275.8 m).

2) No.9 tuff

No.9 sample is the Carboniferous volcanic breccia at the buried depth of 1478.19 m in Well HQ3. Figure 6-7 shows that when the axial differential stress of this rock was less than 165 MPa, the stress-strain curve was appropriately linear, showing elastic deformation. When reaching 165 MPa, the rock started to show plastic deformation, the stress-strain curve bent downward, and the curve slope gradually decreased. The deformation before rock rupture was larger, with an axial strain of more than 1% and hoop strain of more than 0.6%. No.9 sample showed the elastic-plastic deformation characteristics under the confining pressure of 19.36 MPa.

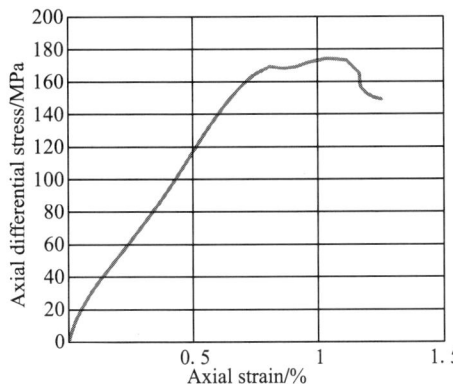
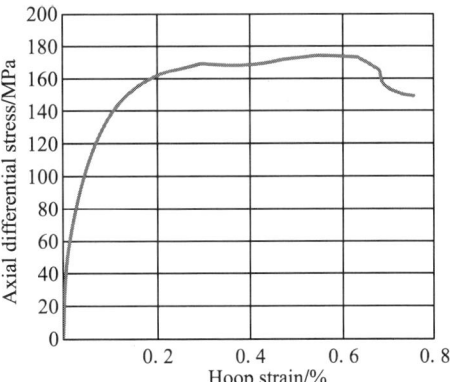

Figure 6-7 Stress-strain relationship of No.9 sample (Confining pressure: 19.36 MPa, HQ3, Carboniferous, tuff, 1478.19 m).

3) No.14 tuff

No.14 sample is the Carboniferous tuff at the buried depth of 2796 m in Well HQ3. Figure 6-8 stress-strain curve shows that when the axial differential stress was less than 250 MPa, the sample showed elastic deformation characteristics; when the axial stress continued to increase to the ultimate strength of 271.35 MPa, the rock stress-strain curve bent downward, showing plastic deformation. The deformation of No.14 sample before fracture was larger, with an axial strain of more than 1% and hoop differential strain of more than 0.8%. On the whole the sample showed elastic-plastic deformation characteristics.

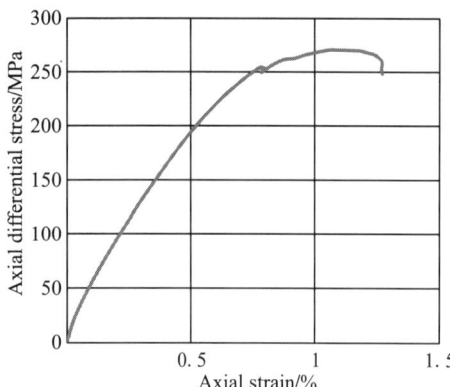
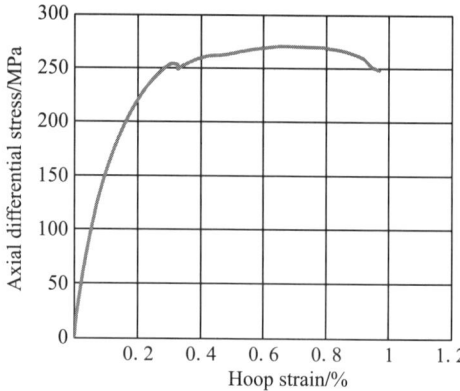

Figure 6-8 Stress-strain relationship of No.14 sample (Confining pressure: 39.7 MPa, HQ3, Carboniferous, tuff, 2796 m).

3. Volcanic breccia

1) No.13 volcanic breccia

No.13 sample is the Carboniferous volcanic breccias at the buried depth of 1210.3 m in HQ3. Figure 6-9 stress-strain curve shows that the rock ultimate strength was 164.13 MPa, when the axial differential stress was less than 164.13 MPa, the rock sample showed elastic deformation; when the axial differential stress was larger than 164.13 MPa, the yield stress and the rock ultimate strength nearly overlapped, and the rock almost showed no plastic deformation before fracture. When the rock ultimate strength reached 164.13 MPa, the total axial strain was 0.58% and total hoop strain was 0.16%. The stress-strain curve reflects typical elastic deformation characteristics.

2) No.4 volcanic breccia

No.4 sample is the Carboniferous volcanic breccias at the buried depth of 1436.11 m. Figure 6-10 stress-strain curve of this sample shows that when the axial differential stress was less than 150 MPa, elastic deformation occurred. During the process when the axial differential stress continued to increase to 153.73 MPa, the stress-strain curve bent downward, the slope gradually became smaller, and the rock sample showed plastic deformation. The deformation of No.4 sample was larger in the whole process of rock mechanics test, with an axial strain of more than 1% and hoop strain of close to 0.8%. On the whole the sample showed elastic-plastic deformation characteristics.

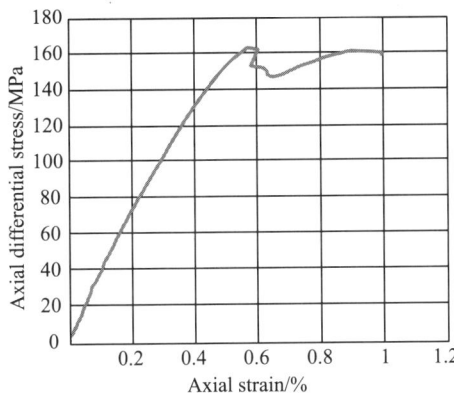
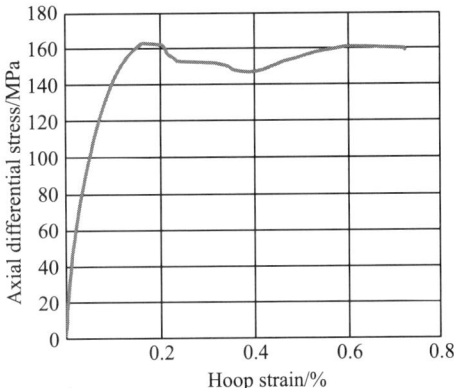

Figure 6-9　Stress-strain relationship of No.13 sample (Confining pressure: 15.61 MPa, HQ3, Carboniferous, volcanic breccia, 1210.3 m).

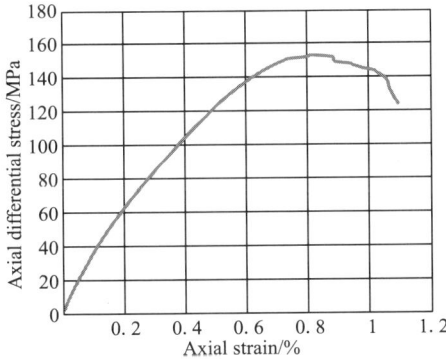
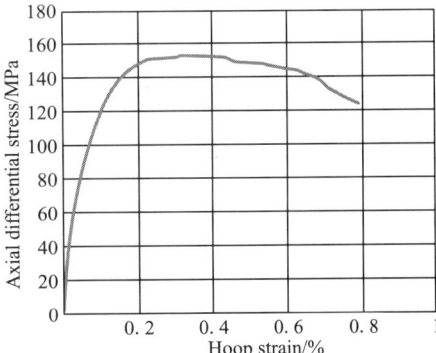

Figure 6-10　Stress-strain relationship of No.4 sample (Confining pressure: 18.96 MPa, HQ101 well, Carboniferous, volcanic breccia, 1436.11 m).

6.1.4　Rock mechanics parameters

1. Tensile strength parameter in the uniaxial test

The tensile strength of rock means the ultimate strength of rock against tension. Generally rock is weak under tension and its tensile strength is usually only 1/10–1/20 of its compressive strength. Rock tensile strength is sensitive to the pores and fractures inside rock, and generally will significantly reduce if there are micro-fractures and pores inside the rock. Besides internal defects, the internal composition of rock also has influence on its tensile strength, such as mineral composition of rock, contact relation between grains, and cementation type, etc.

In terms of the distribution range of tensile strength of the test samples in this research (Figure 6-11), the Permian mudstone samples have a wider range of tensile strength from 4.46 Mpa to 9.51 MPa, the tuff samples have a tensile strength ranging from 9.01 MPa to 13.23 MPa, and volcanic breccia samples of 6.68 MPa. In general, the tensile strength of volcanic rock is higher than that of clastic rock, and is closely related to the density of the tested rock sample. That is to say, the tighter tuff has higher tensile strength than volcanic breccia and mudstone.

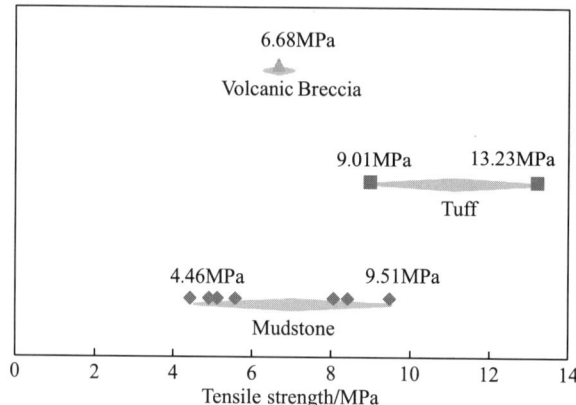

Figure 6-11   Range of tensile strength of different Carboniferous-Permian rocks in the Hala' alate Mountain area.

2. Rock mechanics parameters in the triaxial test

Due to the influence of different rock composition and structure, the variation range of rock mechanics parameters of the tested rock samples in the triaxial test in this research is relatively wide, with the rock compressive strength of around 51.93–475.13 MPa, Young's modulus of deformation of around 18.75–55.92 GPa and Poisson's ratio of around 0.090–0.274. Statistics of mechanical parameters of all test samples has been conducted according to rock types (Table 6-4). From the perspective of rock type, Permian mudstone samples in the research area have higher average compressive strength, average modulus of deformation and average Poisson's ratio than Carboniferous tuff and volcanic breccia samples (Table 6-4).

Table 6-4   Comparison of average rock mechanics parameters of different types of rock samples.

Rock type	Compressive strength/MPa		Modulus of deformation /GPa		Poisson's ratio	
	Range	Average value	Range	Average value	Range	Average value
Mudstone	51.96–475.13	277.27	18.8–55.9	44.1	0.248–0.343	0.279
Tuff	59.68–271.35	190.54	23.6–44.1	38.3	0.09–0.261	0.174
Volcanic breccia	65.13–164.13	127.66	29.4–46.5	37.23	0.16–0.178	0.166

It is found through analysis of the relationship between compressive strength, modulus of deformation and Poisson's ratio that the confining pressure of tested rock samples is closely related to the compressive strength, rock type and Poisson's ratio.

Generally, the rock compressive strength increases with the increase of its confining pressure. It can be seen from the test results of tuff samples, there is a clear linear relationship between confining pressure and compressive strength (Figure 6-12), that is to say, the compressive strength of tuff increases with the increase of confining pressure, while the capacity of fracturing decreases.

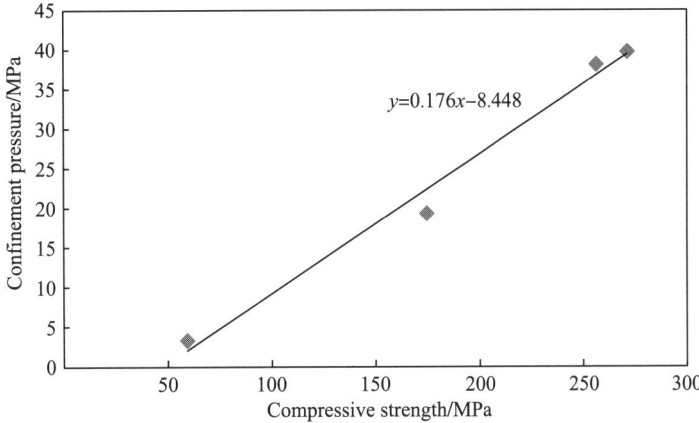

Figure 6-12   Relationship between confining pressure and compressive strength of tuff.

The mineral composition and structural characteristics of rock have influence on its Poisson's ratio. Poisson's ratio of rock mainly reflects the hoop deformation of rock during compression and deformation, and the larger the Poisson's ratio, the larger the hoop deformation. Generally, the larger the Poisson's ratio, the weaker the rock will be during compression and deformation, and thus it will not form fracture easily. From the rock types in this test, the mudstone sampes have the widest range of Poisson's ratio of 0.248–0.343, followed by the tuff samples (0.09–0.261), and the volcanic breccia samples (0.16–0.178). This indicates that, the sequence of difficulty level for fracturing of various rock types in Carboniferous-Permian system is mudstone → tuff → volcanic breccia.

6.2 Factors affecting mechanical properties of rock

For the rock under certain buried conditions, its mechanical properties, besides the internal factors such as rock composition, structure, number of pores and fractures and distribution, are also influenced by the formation conditions to a various extent, the most important of which include temperature, confining pressure, fluid saturation condition and direction of bedded rock samples. Domestic and foreign scholars have conducted a large number of fruitful experiments and discussions on the influence of temperature and confining pressure on the mechanical properties of rock (Ding et al., 2012; Qin et al., 2008; Laiet al., 2004) and results obtained so far mainly includes: both rock strength and Young's modulus of deformation decrease as temperature increases; and rock strength, Young's modulus of deformation and Poisson's ratio all correspondingly increase as confining pressure increases, but the influence degree of temperature and confining pressure on the mechanical properties of rock are different. Handin et al. conducted tests on strength of various sedimentary rocks with depth in underground temperature and confining pressure, the results of which indicated that, less than 5 km deep, the compressive strength of all rocks except salt rock were higher under formation condition than under atmospheric environment condition. Thus, in the shallow strata of earth's crust, the confining pressure has stronger effect on rock strength than temperature. Therefore, it can be considered that, within the depth range of oil and gas reservoir in Carboniferous-Permian system in the Hala'alate Mountain area, temperature has smaller influence on the mechanical properties of rock. In view of the above reasons, this research put emphasis on the influence of rock types and pre-existing weak planes on the mechanical properties of rock.

6.2.1 Influence of rock types on mechanical properties

1. Factors affecting mechanical properties of mudstone

The rock type involves internal factors of rock such as rock composition, structure and number of pores and fractures. For the Permian mudstone in the Hala'alate Mountain area, the relative amount of mud and clastic constituents and the porosity of a rock sample all can influence its mechanical properties. It is found through core observation that the clastic constituents in Permian mudstone contain brittle mineral quartz, which can improve the compressive strength and Young's modulus of deformation, but has little effect on Poisson's ratio. The increase of pores and fractures can reduce tensile strength, Young's modulus of deformation and Poisson's ratio to various extents, but this factor has little influence, because there are limited primary pores in the mudstone. As far as mudstone is concerned, the composition of mudstone itself has key influence on its strength.

2. Factors affecting mechanical properties of volcanic rock

The Carboniferous system in the Hala'alate Mountain area is dense volcanic rock formation. Compared with the Permian mudstone in the research area, Carboniferous volcanic rock has higher density and brittleness, and therefore is easier to fracture under the same tectonic stress condition. It is found by comparing the main volcanic rock types in the area that, the tuff with lower primary pore abundance has higher compressive strength and modulus of deformation than the volcanic breccia with higher primary pore abundance in the triaxial rock mechanics test (Figure 6-13). Thus it can be seen that rock compact degree is an important internal controlling factor for fracture development of Carboniferous volcanic rock in the Hala'alate Mountain area.

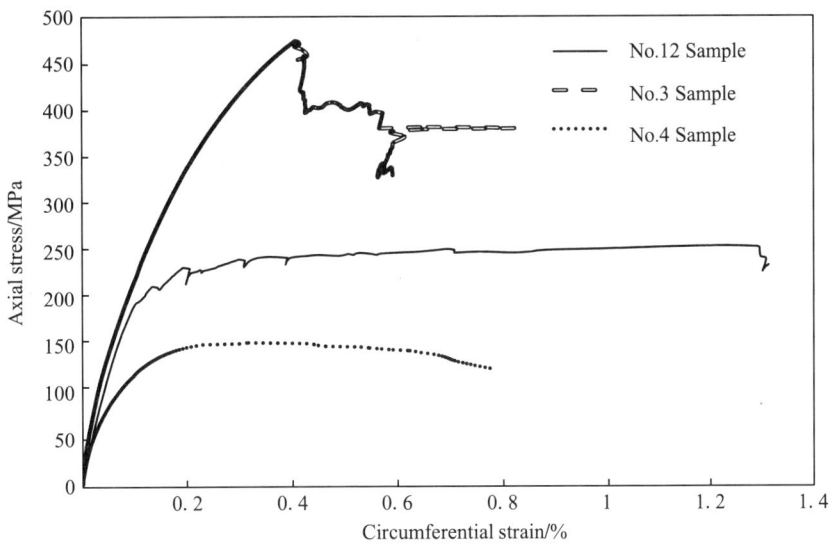

Figure 6-13 Stress-strain relationship comparison of typical rocks in Carboniferous-Permian system.

6.2.2 Influence of pre-existing weak planes on mechanical properties of rock

There are always some "defects" or weak planes of various reasons inside the rock, and the weak planes may be earlier fractures, layers or bedding plane or composition interface. When these structural weak planes exist in the samples, they may significantly weaken the ability of rock against external stress, i.e., reduce their compressive strength and Young's modulus of deformation. For example, No.5 mudstone sample of HQ101 in the Hala'alate Mountain area contains dense network fractures (Figure 6-14), which form weak planes obliquely cross with the axial direction of rock mechanics test samples. As stress was applied to a certain extent, the tested sample firstly fractured along these pre-existing weak planes, but the fractures didn't run through the whole rock sample and other parts were still in good condition. Though the rock sample had certain bearing capacity, as the load increasd and slide along the fracture planes continued, both axial strain and hoop strain increased (Figure 6-15), resulting in significant reduction of both compressive strength and Young's modulus of deformation. While for No.3 mudstone sample, due to the underdevelopment of fractures (Figure 6-14), the pre-existing weak planes has little influence on rock strength and therefore its ultimate compressive strength is higher (Figure 6-15).

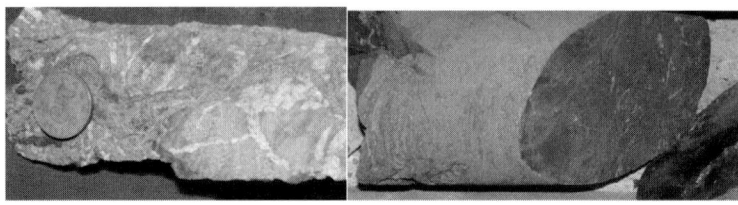

HQ101 in the Hala'alate Mountain area, Permian system, No.5 sample, 2224.1 m.

Well 1 in the Hala'alate Mountain area, Permian system, No.3 sample, 2097.40 m.

Figure 6-14 Comparison of fracture characteristics of sampling section of mudstone in HQ101 and Well 1 in the Hala'alate Mountain area.

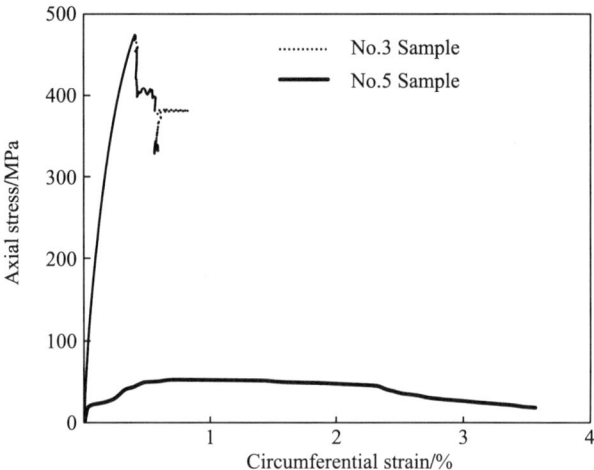

Figure 6-15 Comparison of stress-strain curve of No.5 fracture-developed mudstone and No.3 fracture-underdeveloped mudstone.

6.2.3 Influence of confining pressure on mechanical properties of rock

The rock underground inevitably bears the formation pressure from surrounding rocks—confining pressure. Previous researches (Meng et al., 2000) have proved through a large number of rock mechanics tests that as the confining pressure increases, both compressive strength and Young's modulus of deformation of rock will also increase. That is to say, for the same rock type, the deeper the buried depth of the rock, the larger its compressive strength. It is found through analysis that, under different buried depths, the compressive strength of the Carboniferous tuff in the Hala'alate Mountain area varies widely. It is found through comparison of rock mechanics test results (Figure 6-16) that there is a clear positive correlation between compressive strength and buried depth of rock. Among the 4 tuff samples used in the comparative analysis, No.12 and No.14 rock samples have the maximum buried depth, with a compressive strength of 371.35 MPa and 256.36 MPa respectively; No.9 rock sample with medium buried depth, has a compressive strength of 174.76 MPa; and No.2 rock sample with the shallowest buried depth has the minimum compressive strength of only 59.68 MPa. From the influence of confining pressure on the mechanical properties of rock, for the same rock type, the shallower the buried depth is, the lower the compressive strength and modulus of deformation, and the greater possibility of fracture creation will be.

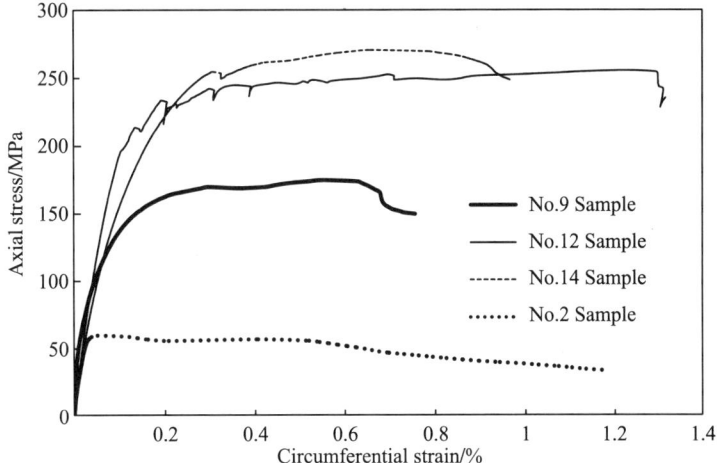

Figure 6-16 Comparison of rock strength of Carboniferous tuff in the Hala'alate Mountain area under different confining pressures (buried depth).

Chapter 7
Factors Affecting Fracture Development and Effectiveness

With underdeveloped primary pores and poor connectivity, volcanic rock is poor in primary quality, and would not become effective reservoir without later reformation. Therefore, later reformation is of great importance for the development of volcanic reservoir, in which the formation of structural fractures takes an important position (Zheng et al., 2012; Dai et al., 2007). Besides increasing the porosity of volcanic reservoir, structural fracture enhances the seepage capacity of formation fluid significantly, making it possible for volcanic rock of low porosity and low permeability to become high quality reservoir. For Carboniferous-Permian reservoirs in the Hala'alate Mountain area, structural fracture not only controls the distribution of high quality reservoirs, but also influences oil and gas enrichment pattern in this area to some extent. Therefore, finding out factor affecting the development of Carboniferous-Permian structural fracture in the Hala'alate Mountain area is of guiding significance for the exploration and development of volcanic oil-gas reservoirs in this area.

7.1 Factors affecting fracture development

The discussion of influence factors of Carboniferous-Permian structural fracture development in the Hala'alate Mountain area is mainly based on the research of rock-mechanics properties, and from three aspects: lithology, stratum thickness and position of external structural stress.

7.1.1 Influence of lithology

Lithologic factors influencing the development of structural fractures include rock components, grain size, cementation condition, content of brittle minerals, etc. These intrinsic factors of rock directly dictate mechanical properties of rocks including compressive strength, tensile strength and shear strength, and thus dictate the difficulty degree of breaking the rock (Fan et al., 2012). Therefore, under the same structural stress, lithologic factors will cause differences in fracture development degree between different types of rocks. Analysis of the experimental results of triaxial rock-mechanic parameter test of Carboniferous-Permian reservoir rock samples in the Hala'alate Mountain area (Table 6-2) shows that, Permian clastic rock has higher compressive strength and deformation modulus than Carboniferous volcanic rock. The analysis of relationship between rock type and compressive stress shows the obtained average compressive stress and average poisson ratio of the three types of rock in a descending order is: mudstone→tuff→volcanic breccia (Figure 7-1). That's to say, under the same structural stress, volcanic breccia is most likely to be broken to form fractures, tuff is in the second place, and mudstone is the most difficult to be broken to form fractures.

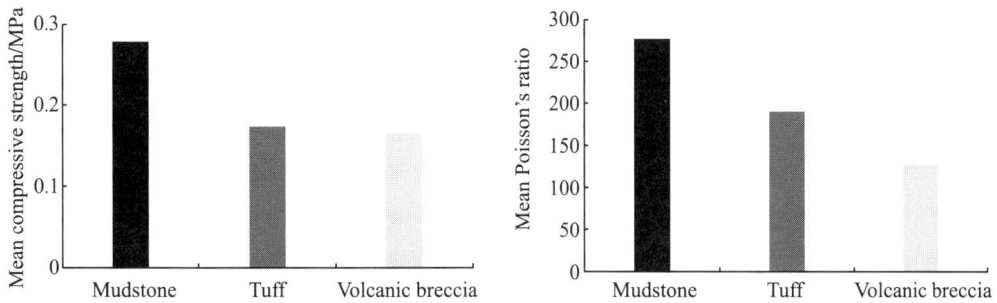

Figure 7-1　Average compressive stress and average Poisson ratio of three types of rock.

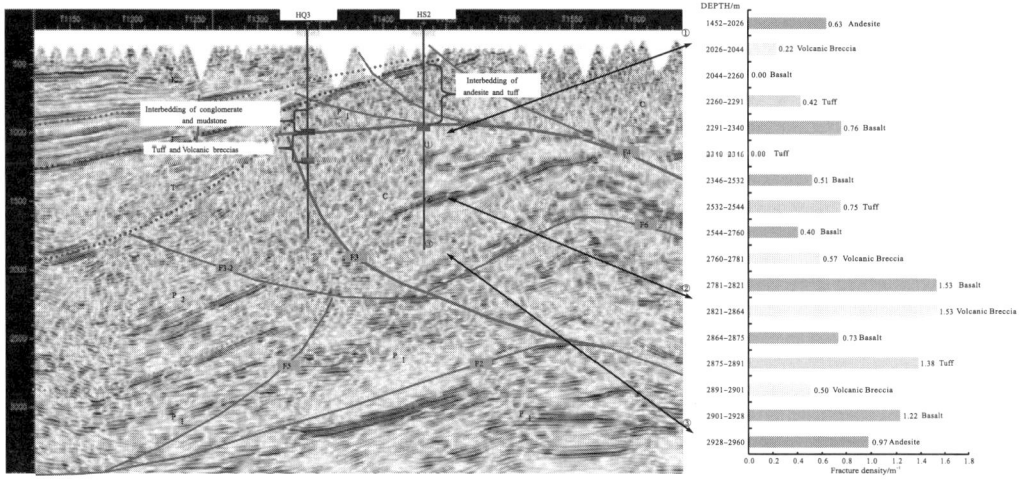

Figure 7-2　Statisticson fracture density of each lithologic section in Well HS2 (from imaging logging data).

As the difficulty degree of rock to be fractured mainly depends on the compact degree and brittle mineral content of the rock, mudstone, a kind of sedimentary clastic rock, has some primary pores and large amount of mud components, and thus lower compact degree and brittle mineral content than volcanic rock, thus higher plasticity, therefore, mudstone is hard to be broken to form fractures. For volcanic rocks, under normal conditions, the more compacted the rock is, the higher its strength, and the more difficult to break it will be. Both tuff and volcanic breccia consist of tephra, the difference is that volcanic breccia has much coarser clastic particles, so it is looser and lower in strength. Statistics on relationship between fracture development density and lithology from imaging logging section of Well HS2 (Figure 7-2) reveal that, volcanic breccia has much higher fracture density than tuff. Besides fracture development position, formation thickness, confining pressure of formation, etc. the intrinsic reason causing high fracture density in volcanic breccia is its compact degree.

7.1.2 Influence of formation thickness

The control of stratum thicknesson fracture development mainly manifests in the following aspects: fracture development and distribution is controlled by stratum thickness and fractures only develop in stratum and end at lithologic interface. On one hand, thickness of single layer has obvious control on the degree of fracture development, that's to say, fractures are more developed in thin strata; on the other hand, the thinner the stratum, the finer the clastic particles, the more compacted the stratum will be, and the more developed fractures will be (Zhou et al., 2003).

Previous studies indicate that, within a certain scope of stratum thickness, there is a negative correlation between stratum thickness and fracture density. That's to say, the thinner the stratum, the more developed the fractures will be. Because of the poor stratification, the volcanic rock in the research area, is mainly in compacted massive beddings, so stratum thickness may not have obvious influence on fracture development in Carboniferous volcanic rock and volcaniclastic rock.

7.1.3 Influence of structural position

Density of reservoir fractures is closely related to structural position, thus structural position is an important factor influencing the development and distribution of reservoir fractures. Stress field strength in different structural positions differ widely, making fracture density obviously different correspondingly. It is generally believed that the position with large structural curvature change is most favorable for forming structural fractures (Cao et al., 2007; Zhao et al., 2005; Wang et al., 2002.).

Permian-Carboniferous in the research area are abundant in fractures. Existing research results indicate that the development degree of structural fractures is closely related to fault activities. Fault control on the development and distribution of fractures by is mainly realized through controlling the distribution of local tectonic stress nearby. For example, from seismic section of Well HS2 (Figure 6-2) it can be seen that, strata cut through by faults have more structural fractures, and much higher fracture density than strata with no fault passing from

Chapter 7 Factors Affecting Fracture Development and Effectiveness

fractures statistics by imaging logging. In those parts where tectonism is intense, the influence of structure to fracture development may even cover the negative influence of lithology.

Statistics of fracture density of five wells (Well HS1, HS2, HQ3, HQ6, HQ101) in the Hala'alate Mountain area (Figure 4-3) show that, Well HS2 and HS1 Well in the northernmost research area have the largest fracture density of 0.56 m^{-1} and 0.53 m^{-1} respectively. While Well HQ3, HQ6, and HQ101 in the south have the largest fracture density of no more than 0.22 m^{-1}. Single-well fracture density in the Hala'alate Mountain area has the feature of "higher in the north and lower in the south", which is because in tectonic evolution period, this area experienced compressive stress coming from north-west direction.

Statistics of single-well fracture density of five wells (HS1, HS2, HQ3, HQ6, HQ101) in the Hala'alate Mountain area (Figure 7-3) show that, the northernmost Well HS2 and HS1 in the research area have the largest fracture density of 0.56 m^{-1} and 0.56 m^{-1} respectively. While Well HQ3, HQ6, and HQ101 in the south have a fracture density of less than 0.22 m^{-1}. Single-well fracture density in the Hala'alate Mountain area is higher in the north and lower in the south, because in structural evolution period, this area experienced compressive stress coming from north-west direction, which reflects that structural activity from this direction was more intense.

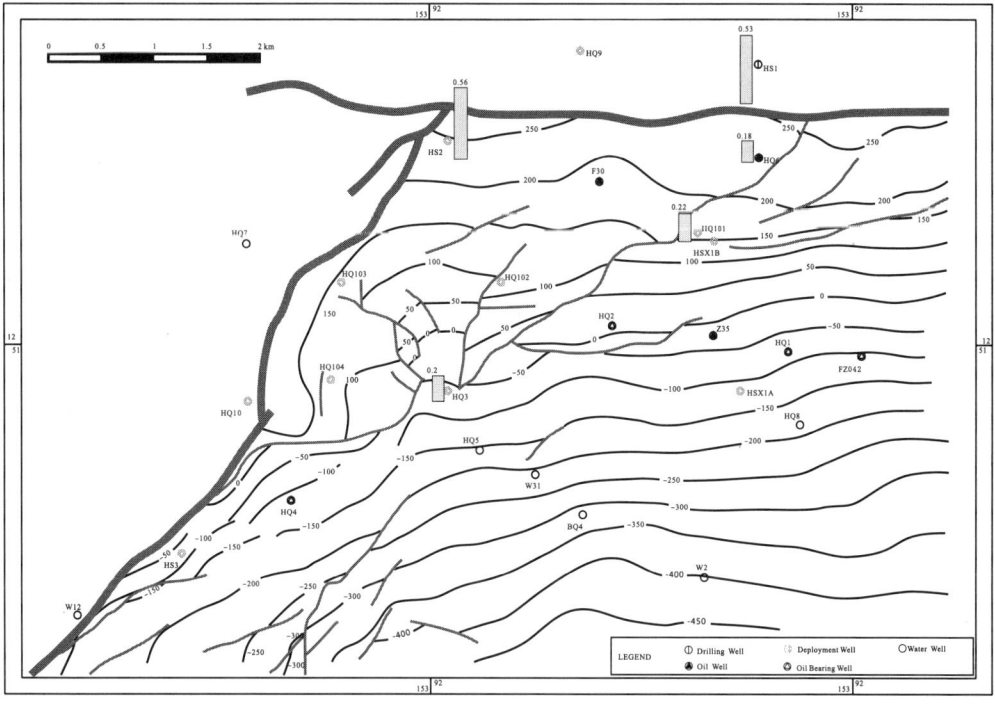

Figure 7-3 Relationship between single-well fracture density and structural position in the Hala'alate Mountain area.

7.2 Development stages of fractures

The Hala'alate Mountain work area is situated in the west of Hala'alate Mountain at the northwest margin of Junggar Basin, and is also a piedmont thrust belt belonging to Wuxia fault belt. Since experiencing complicated and multi-stage tectonic movements, there develop structural fractures widely in Carboniferous and Permian strata in this area. To find out the genetic mechanism and development stage of Carboniferous-Permian structural fractures in this area, it is necessary to incorporate regional tectonic evolution of Junggar Basin into the fracture study to get a clear understanding on the development stages of structural fractures in the whole research area under the specific tectonic movement background.

7.2.1 Sequence of fracture development

Through core observation and statistics on high-resistance fractures and high-conductivity fractures in Hala'alate Mountain work area in imaging logging, and in combination with seismic sections interpreted, we have analyzed the relationship between reverse faults in Carboniferous-Permian and characteristics and stages of fracture development.

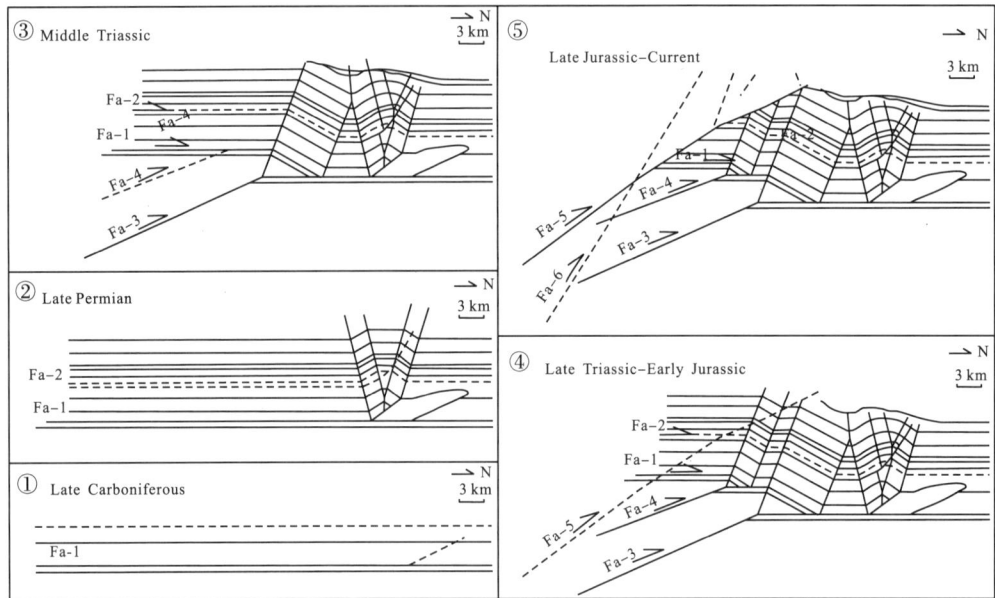

Figure 7-4　Formation and evolution model of the west nappe in Hala'alate Mountain (drawing) (Liu, 2012).

Seismic sections interpreted in Hala'alate Mountain work area reveal that there are five larger faults and the associated induced-faults in Carboniferous-Permian, namely, F2, F3, F4,

F5 and F6. It is generally thought that near-fault areas are places where concentrated tectonic stress releases, the fracture density there is obviously higher than the strats with no fault passing through or far away from faults, moreover, because faults have certain occurrence, structural fractures formed by stress release of the fault also have certain occurrence, that's to say, fracutes controlled by different faults have different occurrences. On this basis, combining the analysis of mutual cutting relationship of fractures and the analysis of fracture development in imaging logging, the sequence and stages of fracture development can be figured out.

Through the analysis of seismic section and tectonic evolution of the Hala'alate Mountain work area, it is concluded that there are five larger faults and they have an obvious sequence of movement in this area. From the figure of formation and evolution model of the west nappe of Hala'alate Mountain (Figure 7-4), it can be figured out that the active sequence of reverse faults in the Hala'alate Mountain work area is: F2→F6→F3-F5→F4. Combining core observation of drilled wells, statistics on fractures in imaging logging and interpretation result of seismic sections, the stage of structural fractures in this region have been sorted out carefully.

1.Sequence of fracture development in typical drilled wells

1) Well HS1

Intersecting F3 and F4 reverse faults, and close to F6 reverse fault, Well HS1 is located in a position with fairly strong tectonic deformation, so fractures are quite developed in Carboniferous and Permian there. Structural fracture parameters obtained from imaging logging of Well HS1 reveal that there exist two sets of fractures of obviously different dominant dip in Permian mudstone at about 2152 m depth. Fractures in mudstone section at 1960-2152 m logging depth dip dominantly south-east, with dip angle concentrating in the range of 40°-50°; while fractures in mudstone section of 2152-2432 m logging depth dip dominantly north-west, with dip angle concentrating in the range of 50°-60°. From seismic interpretation section of Well HS1 (Figure 7-5), it can be seen that mudstone section of 1960-2152 m roughly falls into activity area of Fault F6 which dips south-east at low angle at Well HS1, while mudstone section of 2152-2432 m is basically controlled by Fault F3 which dip north-west at high angle. Observation of the core in the sixth core run (2151-2157.7 m) from Well HS1 shows that there are net-shaped fractures (Figure 7-6), which is an obvious result of superposed fault activities of F6 and F3 with different stress characteristics.

Fracture development obtained by imaging logging reveals that south-east dip fractures controlled by Fault F6 are obviously lower in density than north-west dip fractures controlled by Fault F3, and high-resistance fractures mainly exist in the section controlled by Fault F6 (Figure 7-7), which indicates that fractures controlled by Fault F6 were formed earlier, and higher in filling degree than those controlled by Fault F3, so they have lower overall effectiveness than those controlled by Fault F3.

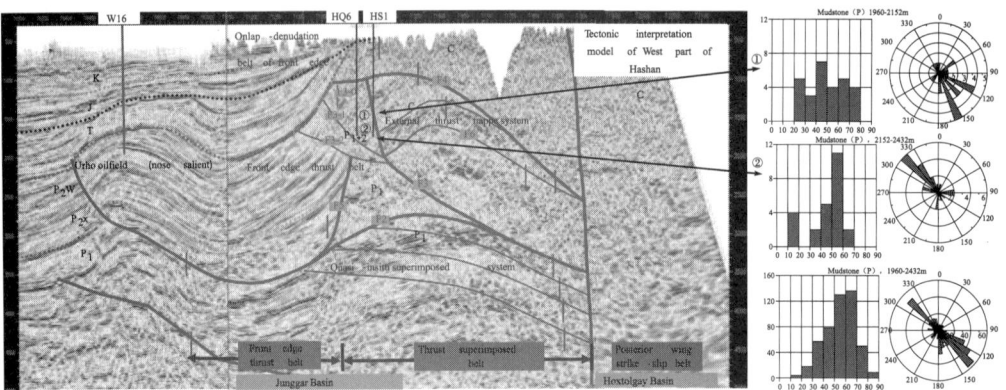

Figure 7-5 Relationship between occurence of high-resistivity fractures in Permian mudstone section and structural location in Well HS1.

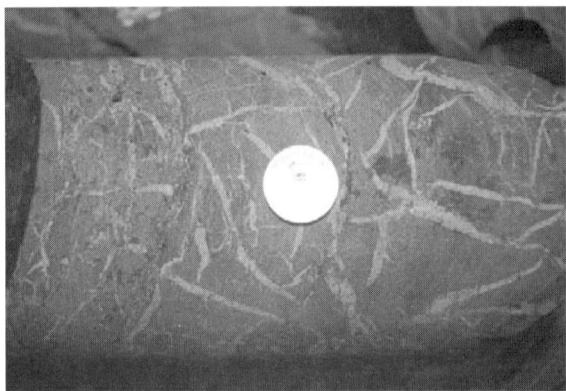

Figure 7-6 Dense net-shaped fractures (Well HS1, mudstone, Permian System, 2153.7 m).

In summary, fracture development in Permian mudstone section of Well HS1 is mainly controlled by the activities of Fault F3 and F6, and fractures controlled by Fault F3 which was active later have higher development degree and effectiveness.

2) Well HQ6

Situated south of Well HS1, Well HQ6 drilled through the shallow F4 reverse fault and into Permain mudstone. From seismic interpretation section through Well HQ6 (Figure 7-8), it can be seen that, there is a distance between Well HQ6 and Fault F3, so it is less affected by Fault F3 than Well HS1.

In shallow formations in Well HQ6, a series of low-angle structural fractures and induced high-angle reverse faults dipping mainly north-west were formed due to the activity of Fault F4. In deep sections of Well HQ6, influenced by activities of Fault F6 in early stage and Fault F3 in later stage, there are fractures of obviously different dip and dip angle. The influence of fault activities in two stages on fracture development in this well is similar to the circumstance in Well HS1. The fractures formed by activity of early Fault F6 dip south-east dominantly, and have higher filling degree and higher resistancein imaging logging because of the earlier forming time (Figure 7-9). In contrast, the fractures formed by activity of late Fault F3 dip dominantly north-west, and low in filling degree, they have high-conductivityin imaging logging (Figure 7-10). Because Fault F3 does not pass through this well, it is its branch that affected Well HQ6, the fractures dipping north-west in this well have smaller dip angle than those in Well HS1. Similarly, there exist net-shaped fractures in cores of well section jointly affected by Fault F6 and Fault F3 (Figure 7-11), but as the structural stress in this well is weaker than in Well HS1, the net-shaped fractures are lower in development degree.

In summary, high-conductivity fractures dipping north-west of good effectiveness in deep formations of Well HQ6 are mainly caused by Fault F3 (branches). In shallow formations, due to the effect of Fault F4 (branches), high-conductivity fractures dipping north-west with good effectiveness were formed too.

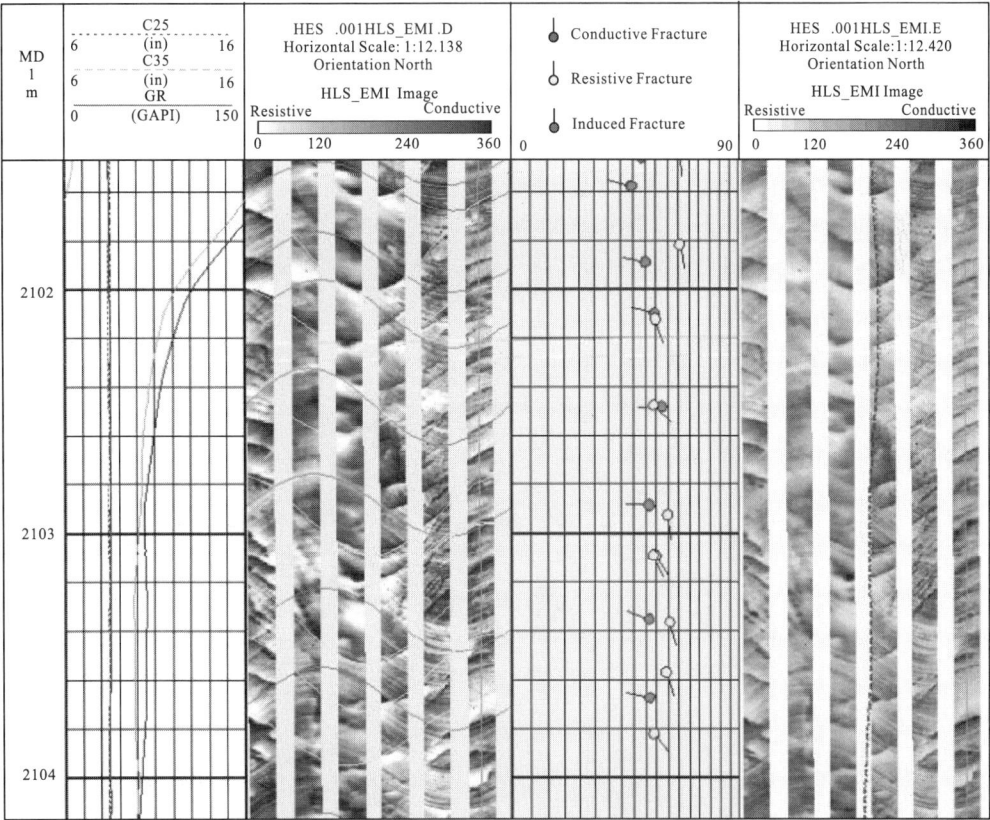

Figure 7-7 Characteristics of high-resistance farctures controlled by Fault F6 in Well HS1 (Well HS1, mudstone section, Permain System, 2102–2104 m).

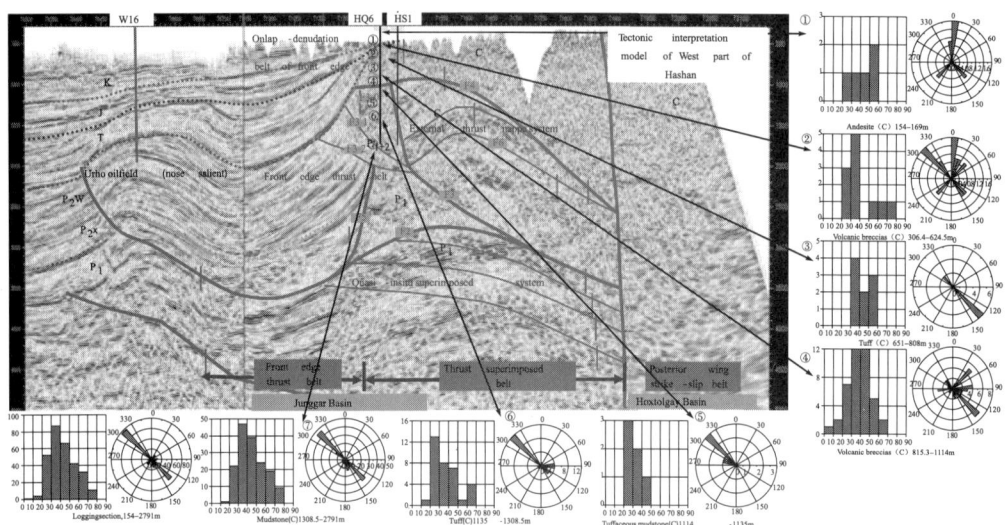

Figure 7-8 Relationship between occurence of high-conductivity fracture and structural location of whole logging section of Well HQ6.

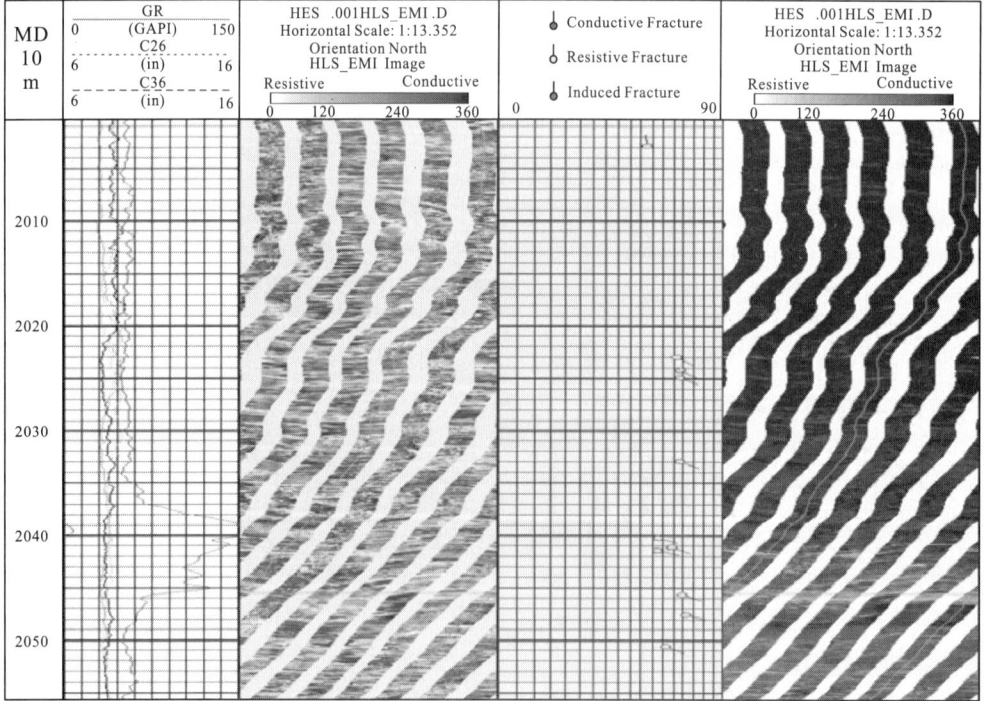

Figure 7-9 Characteristics of high-resistance farctures controlled by Fault F6 in Well HQ6 (Well HQ6, mudstone section, Permain System, 2000–2050 m).

Chapter 7 Factors Affecting Fracture Development and Effectiveness

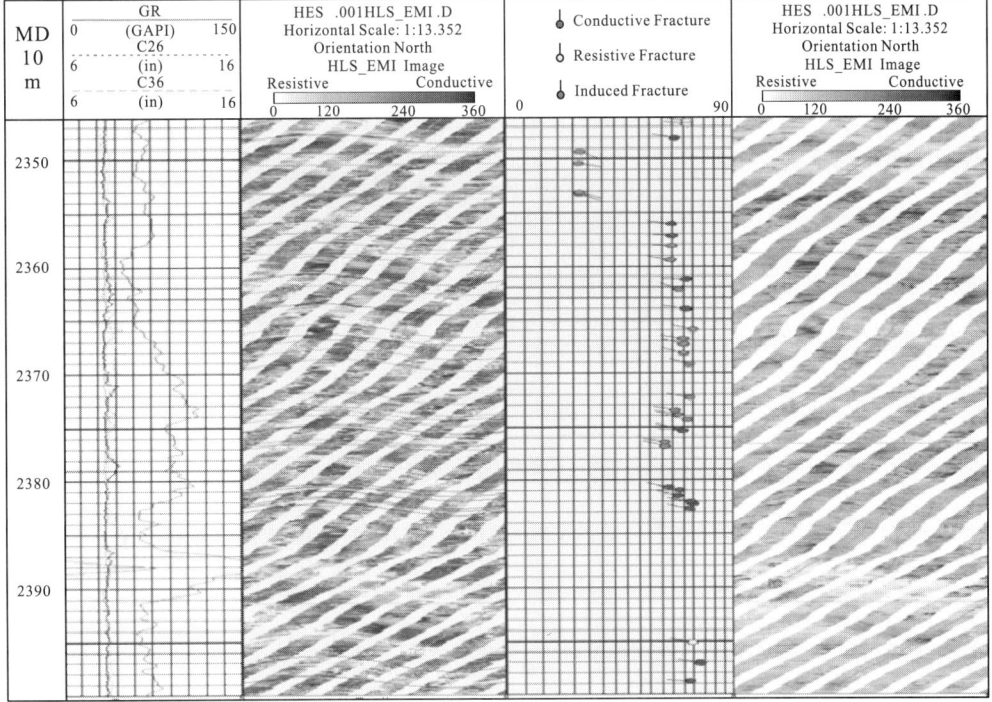

Figure 7-10 Characteristics of high-conductivity farctures controlled by Fault F3 in Well HQ6(Well HQ6, Mudstone Section, Permain System, 2350-2400 m).

Figure 7-11 Net-shaped fractures(Well HQ6, Mudstone, Permian System, 1919.70 m).

3) Well HQ101

Situated southwest of Well HQ6, Well HQ101 has Carboniferous volcanic breccia in the upper section and Permain mudstone in the lower section. Well HQ101 mainly passes through F4 reverse fault, and is close to F3 reverse fault. From interpreted seismic section (Figure 7-12), it can be seen that Well HQ101 is more influenced by F4 reverse fault.

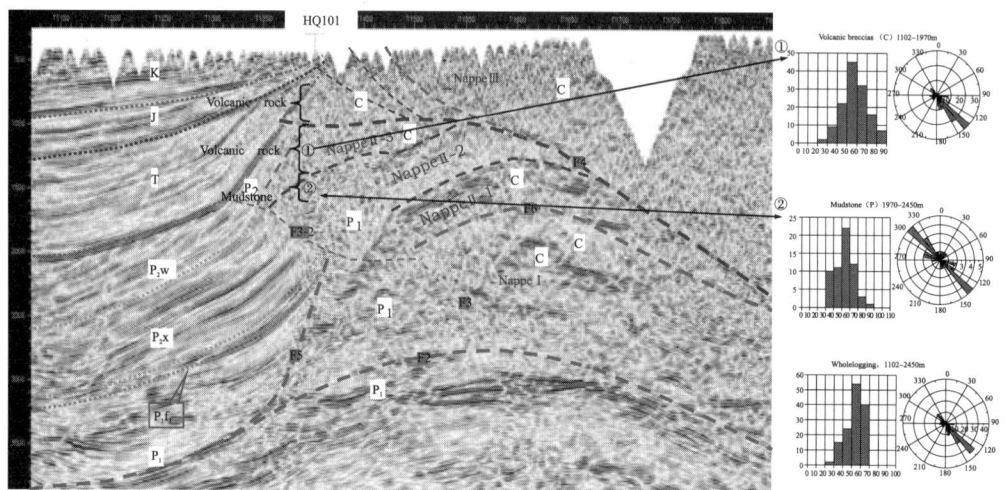

Figure 7-12 Relationship between occurence of high-conductivity fracture and structural location in Well HQ101.

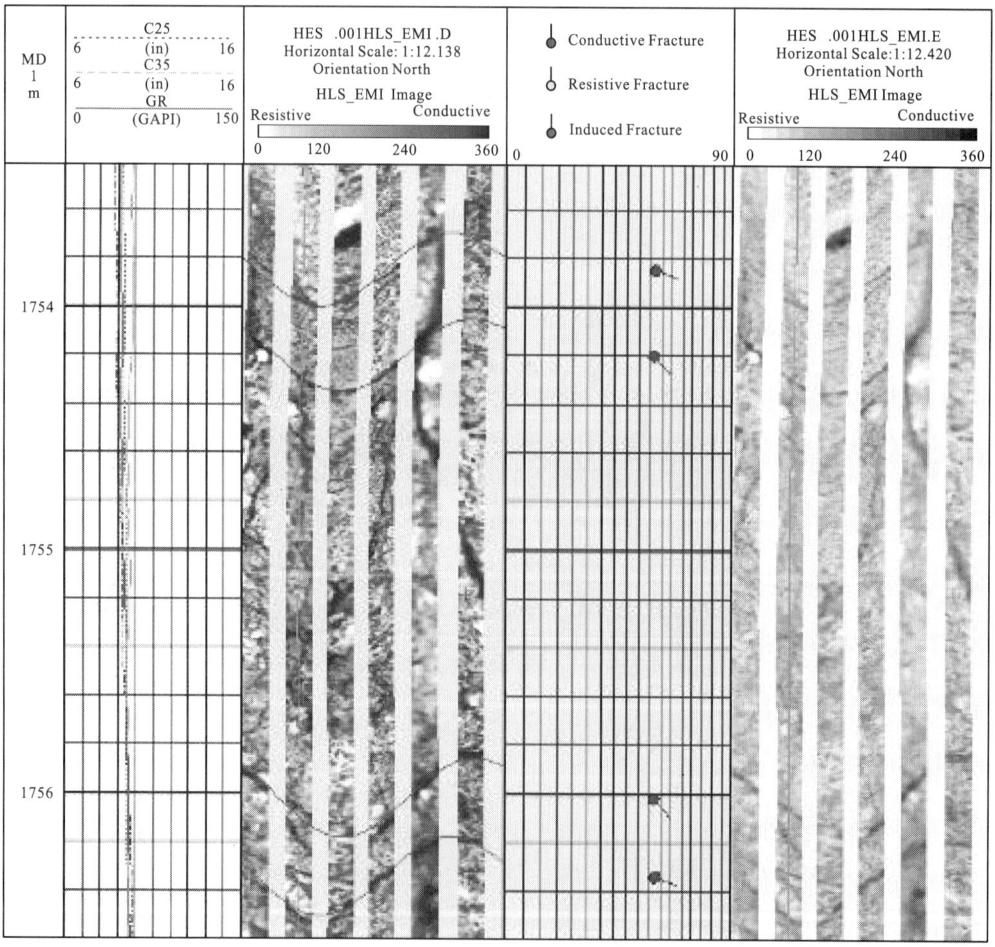

Figure 7-13 Characteristics of high-conductivity farctures controlled by Fault F4 (branches) in Well HQ101 (Well HQ101, Volcanic Breccia Section, Carboniferous System, 1754–1756 m).

High-conductivity fractures dipping south-east related to the activity of F4 reverse fault are very widespread in the shallow Carboniferous volcanic breccia, F4 Fault might induce branch faults of higher dip angle, which in turn caused the appearance of high-conductivity fractures dipping south-east dip at higher dip angle within the controlling scope of Fault F4 (Figure 7-13). In contrast, in deep buried Permian mudstone, the joint effect of Fault F4 and F3 resulted in the formation of south-east and north-west dipping high-conductivity fractures, in which the high-conductivity fractures dipping north-west are larger in number.

Imaging logging reveals that, high-conductivity fractures in Well HQ101 take an absolute majority, which is directly because the structural fractures are mainly controlled by F4 and F3 reverse fault which were active late. Full-filling net-shaped fractures in the mudstone of 2236m are cut through by high angle fractures (Figure 7-14) with oil-immersion in the interface, which indicates that the net-shaped fractures were formed and filled with calcite before the activities of Fault F3 and F4, then crude oil derived from source rock migrated into the fractures, that implies that this kind of fracture is good in effectiveness.

Overall, fracture density in Well HQ101 is smaller than that in Well HS1, but close to that in Well HQ6 (Figure 4-5). Mainly affected by the activities of F3 Fault, high-conductivity fractures dipping north-west formed in late stage take dominance, with fairly high effectiveness.

Figure 7-14 Net-shaped fractures cut through by high angle fractures (Well HQ101, Mudstone, Permian System, 2236.15 m).

4) Well HQ3

Situated south of Well HQ2, Well HQ3 (Figure 7-15) drills through Fault F4 and F3 according to seismic section through this well. Statistics and analysis of fracture development revealed by imaging logging of Well HQ3 show that there mainly develop high angle fractures of north-west dip (Figure 7-16), which are interpreted as high-conductivity fractures from imaging logging.

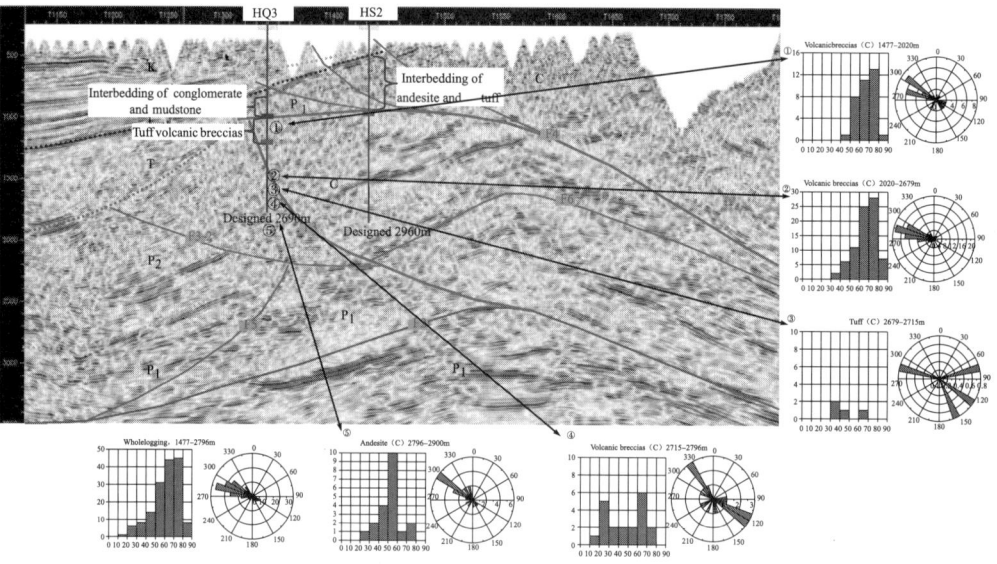

Figure 7-15 Relationship between occurence of high-conductivity fracture and structural location of whole logging section of Well HQ3.

Shallow formations in Well HQ3 are mainly affected by the activity of Fault F4. Because Fault F4 is parallel to rock bed and nearly horizontal at this well, the structural fractures formed by its activities are of low dip angle or nearly horizontal too. Because activity time of F4 Fault is late, the fractures are basically not filled, with oil trace seen in some of them(Figure 7-17).

Figure 7-16 Un-filled high angle fractures(Well HQ3, Carboniferous System, Volcanic breccia, 2643.20 m).

In summary, development degree of high-conductivity fractures in Well HQ3 is similar to that in Well HQ6. In this well, shallow formations are controlled by Fault F4 (branches), and the deep formations by Fault F3. Imaging logging interpretation of Well HQ3 reveals that high-conductivity fractures of north-west dip have absolute dominance (Figure 7-18), which indicates that development of high-conductivity fractures in Well HQ3 is mainly related to the activity of F3 reverse fault.

Figure 7-17 Low angle fractures, with oil trace in fracture surface (Well HQ3, Carboniferous System, Volcanic Breccia, 1475.92 m).

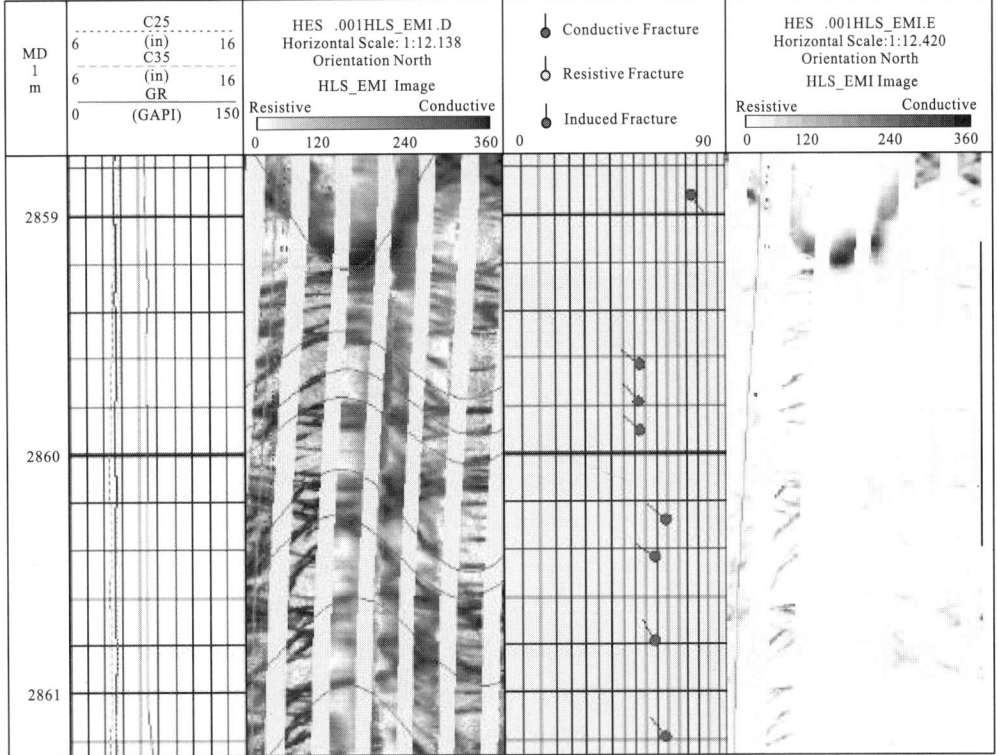

Figure 7-18 Characteristics of high-conductivity farctures controlled by Fault F3 in Well HQ3 (Well HQ3, Andesite Section, Carboniferous System, 2859–2861 m).

5) Well HS2

Situated in the most northwestern part of the study area out of the five analyzed wells, and closer to piedmont fold belt, the place where Well HS2 is has stronger tectonic stress. From seismic interpretation section through Well HS2 (Figure 7-19), Well HS2 mainly goes through F4 overthrust fault and its Carboniferous nappe. Because of the strong intensity of Fault F4 (branches) activity, the fracture density of imaging logging sections of Well HS2 is the largest among the five wells analyzed (Table 4-2).

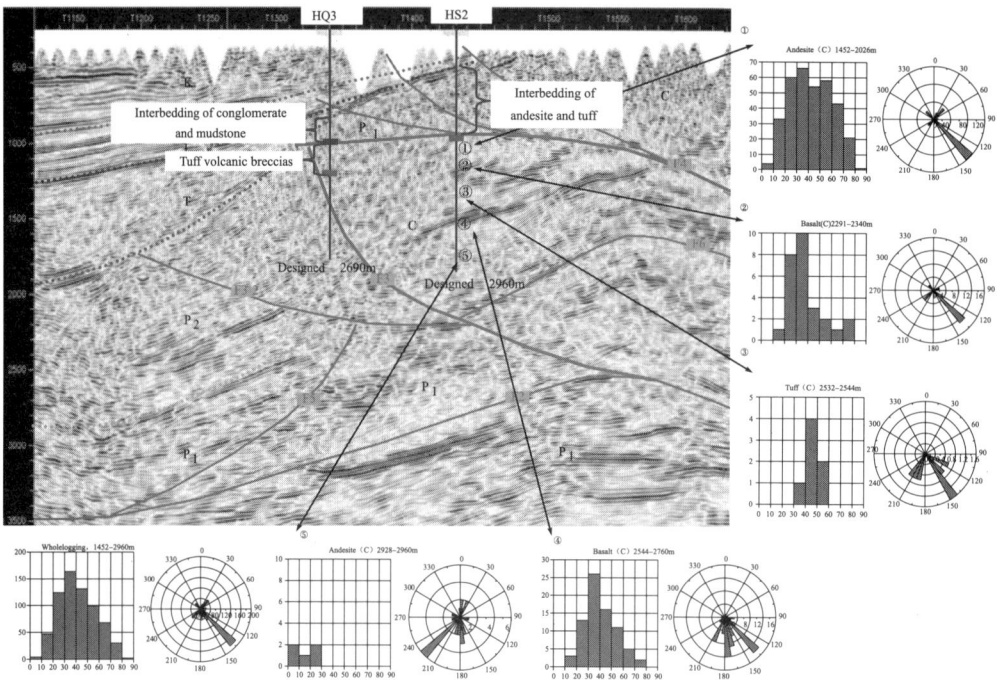

Figure 7-19   Relationship between occurence of high-conductivity fracture and structural location of whole logging section of Well HS2.

In addition, the seismic section through Well HS2 shows that Fault F4 in this well is mainly of south-east dip and medium-low dip angle. Because the activity time of F4 Fault is late, fractures formed in Well HS2 are mainly high-conductivity fractures of southeast dip and medium-low dip angle (Figure 7-20). Core observation shows that these fractures are generally low in filling degree or even not filling (Figure 7-21).

Because Well HS2 is mainly controlled by the activity of Fault F4, structural fractures of southeast dip and medium-low dip angle are abundant in this well. The density of high-conductivity farctures in it is the largest among the five wells analyzed, moreover, the fractures are low in filling degree and high in effectiveness because of late formation time.

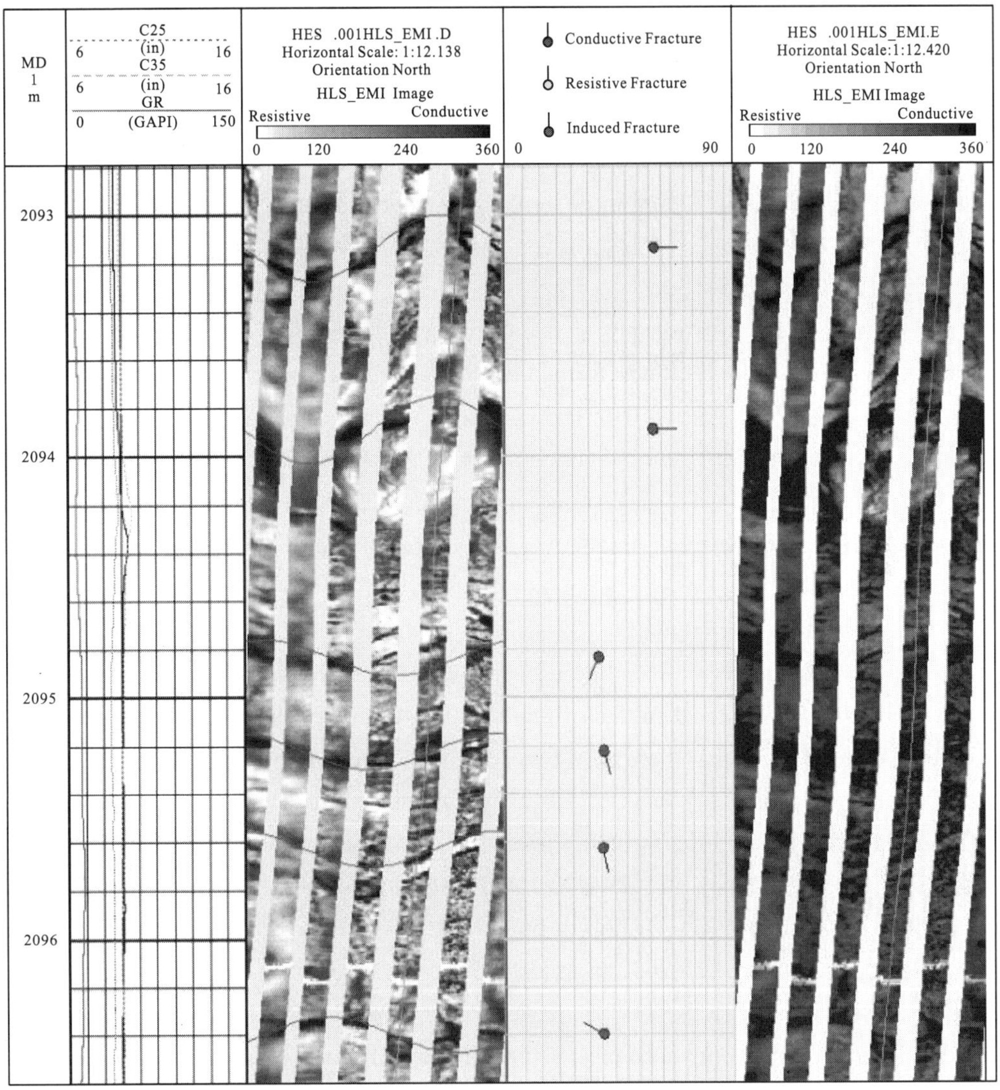

Figure 7-20 Characteristics of high-conductivity farctures controlled by Fault F4 (derived fault) in Well HS2 (Well HQ2, Volcanic breccia section, Carboniferous System, 2093–2096 m).

Figure 7-21 Unfilled structural fractures of medium-low dip angle (Well HS2, Carboniferous System, tuff, 1209.85 m).

2. Development sequence of fractures in the Hala'alate Mountain area

Based on the analysis of development sequence of Carboniferous-Permian fractures in each typical well in the Hala'alate Mountain area, the development sequence of Carboniferous-Permian fractures has been sorted out according to activity sequence of major faults. Activity sequence and development characteristics of major reverse faults in the research area are as follows:

1) Fault F6 and associated fractures

It is inferred from cutting relationship of main reverse faults in the Hala'alate Mountain area (Figure 7-8) that F6 reverse fault was formed earlier than F3 and F4 reverse faults. The seismic section through Well HS1 and HQ101 shows Fault F6 is of south-east dip and medium dip angle. In addition, in the Permian mudstone section of 2000–2050 m in Well HS1 cut through by F6 reverse fault, imaging logging shows the fractures are high-resistance ones with high filling degree (Figure 7-9), which also proves that F6 reverse fault and its associated fractures were formed earlier.

2) Fault F3 and associated fractures

From seismic sections, it can be seen that F3 reverse fault cuts F6 reverse fault, indicating that active time of Fault F3 is later than that of Fault F6. Statistics of fractures revealed by imaging logging of Well HS1 show that there exist two sets of fractures of obviously different dominant dip in Permian mudstone at about 2152 m, in which fractures in the shallower section (1960–2152 m) are mainly dipping south-east with a dip angle in the range of 40°–50°; while fractures in the deeper section (2152–2432 m) mainly dip north-west with a dip angle in the range of 40°–50°. From seismic section through Well HS1 (Figure 7-8), we can see that, the northwest dip fractures in 2152–2432 m mudstone section are mainly controlled by F3 reverse fault. Core observation of 2151–2157.7 m well section in Well HS1 shows that, net-shaped fractures are developed in this section (Figure 7-6), which must be the result of superposed activity of F6 and F3 reverse faults of different structural stress directions successively.

Imaging logging reveals that, fractures of northwest dip controlled by F3 reverse fault are mainly high-condctivity fractures (Figure 7-10), and much higher in density than fractures of southeast dip controlled by F6 reverse fault (Figure 7-9), therefore, in terms of either fluid seepage capacity now or scale, fractures associated with F3 reverse fault are better than those associated with F6 reverse fault.

3) Fault F4 and associated fractures

Situated southwest of Well HQ6, Well HQ101 has Carboniferous volcanic breccia in the upper section, and Permain mudstone in the lower section. Well HQ101 mainly passes through F4 reverse fault, and is close to F3 reverse fault. From the seismic section (Figure 7-12), it can be seen that Well HQ101 is more influenced by F4 reverse fault. Particularly, high-conductivity fractures of south-east dip related to the activity of F4 reverse fault are widespread in the shallow Carboniferous volcanic breccia. F4 Fault also derived branch faults of higher dip angle, causing the appearance of high-conductivity fractures of south-east dip and higher dip angle within the controlling scope of F4 Fault (Figure 7-20). In contrast, in deep buried Permian mudstone, influenced by fault F4 and F3 related faults, there develop high-conductivity fractures of south-east and north-west dip, in which the high-conductivity fractures of north-west dip controlled by Fault F3 take a majority in number.

Imaging logging reveals that, high-conductivity fractures in Well HQ101 take an absolute majority, which are directly related to the late active time of F4 and F3 reverse fault that control the structural fractures there. Full-filling net-shaped fractures in mudstone of 2236m are cut by high angle fractures with oil immersion in the interface (Figure 7-14), which indicates that the net-shaped fractures were formed and filled before the activities of F3 and F4 faults, and the high angle fractures associated with Fault F4 formed in later stage have good effectiveness.

7.2.2 Tectonic evolution and fracture development stage

Fracture development characteristics reflected by cores and imaging logging has been elaborated in the above section, but core observation and imaging logging are just limited in wells, generally, which can not reflect cutting and crossing relationship of fractures in the whole area, not allowing observation of spatial configuration and distribution of fractures of different stages of different stages and their spatial distribution. Therefore, it is difficult to find out formation time of structural fractures and its relationship with oil and gas accumulation. As Carboniferous-Permian fractures in the Hala'alate Mountain area are mainly structural fractures, development stage of fractures can be calibrated through the analysis of relationship between the occurrence of structural fractures and characteristics of tectonic movement of each stage. Since tectonic evolution of the basin is closely related to regional geotectonic evolution, the analysis of tectonic evolution of the Hala'alate Mountain area should be based on tectonic evolution of Junggar Basin (Table 1-1), and be combined with actual tectonic condition of northwest margin of Junggar Basin. According to previous studies (Lin et al., 2006), after the formation of Carboniferous volcanic rock in northwest margin of Junggar Basin, this region has experienced 4 stages of tectonic movements: tectonic compression of middle-late Hercynian movement, successive development of Indosinian movement and external-compression and internal-tension of Yanshan movement, ending in the current basin-mountain coupling relationship and feature of the Hala'alate Mountain area (Figure 7-22).

In middle Hercynian period, nappe tectonic belt in northwest of Junggar Basin had not formed yet, because of collision of plates, the aulacogen between Junggar terrane and Kazakhstan terrane started to close (Xu et al., 2014), under the effect of NW-SE tectonic compression, activity of basement strikeslip faults of northwest trend vertical to tectonic stress of northeast direction appeared in the northwest margin of the area, so while the forming of north east tectonic system prototype, two sets of X-shaped conjugate shear fractures in NW and

NE trend almost vertical to original horizontal rock stratum were formed (Yang et al., 2011). In middle-late episode of middle Hercynian period, persistent NW tectonic stress caused the formation of high-angle horizontal tension fractures of NW direction in the formation. Meanwhile, under the effect of continuous strengthing tectonic stress, the existing NE and NW reverse faults gave rise to a series of associated shear fractures.

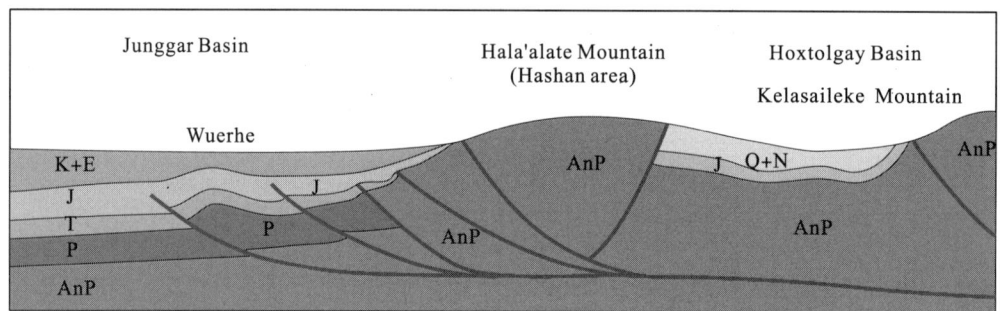

Figure 7-22 Current basin-mountain coupling relationship of the Hala'alate Mountain area (according to Feng Jianwei, modified) (Feng, 2008).

In late Hercynian period, under the effect of NW tectonic stress once again, on the basis of existing NE compresso-crushed zone, NW margin of Junggar Basin continuously developed into a large-scale NE reverse fault and experiened sinistral rotational shear deformation to some extent (Tan et al., 2008). The tectonic movement of this stage reformed the two sets of X-shaped conjugate shear fractures of NE and NW directions formed in earlier stage, making them show characteristics of "shear before tension". Because of the continuous effect of NW tectonic stress, on the basis of original fractures, two sets of X-shaped and high-angle conjugate shear fractures of NE direction formed in the formations, and the scale of high-angle horizontal tension fractures formed in earlier stage vertical to the direction of tectonic principal stress were enlarged further (Feng, 2008). Because of large development scale, wide width and long extension, conjugate shear fractures of late Hercynian period are the most extensive type of structural fracture in the research area.

Indosinian period was the time when tectonic framework of Junggar Basin basically finalized (Zhu et al., 2015; Tan et al., 2008; Lu et al., 2008), and also an important period for the formation of the Hala'alate Mountain nappe. In this tectonic stage, intensity of tectonic movement of northwest margin of Junggar Basin was weaker to some extent than late Hercynian period, but activities of thrust-nappe structures of Wuxia fault belt did not stop and continued to move towards the inner basin. In this process, because of imbalance of nappe stress, Baiyanghe basement fault was active again, and cut all Permian strata. (Feng, 2008) Daerbut fault in the north of Hala'alate Mountain showed characteristics of sinistral strike-slip activities in the process of Wuxia fault belt multi-stage episodic tectonic movements, forming a series of sinistral echelon fracture groups of NWW direction (Fan et al., 2014). In late Indosinian period, short-term and strong soft collision occurred between Junggar terrane and Kazakhstan terrane, at this time, Indian plate to the south subducted towards north strongly, leading to northward compression of Junggar terrane in triangle shape. In the meantime, under the effect of reverse tectonic stress of Siberian plate towards south and the blocking of thick Permian-Triassic strata deposited in the basin front, large scale of overthrusts and nappes occurred in Western Junggar Basin (Feng, 2008). As geosuture of plate margin, Daerbut fault moved vertically and backward towards north and in this process tectonic stress was released, leading to the formation of recoil faults and associated activities of strike-slip faults in Daerbut north area. Under the effect of reverse nappe tectonic activity, Carboniferous between these two faults was uplifted, causing the basic features of current Hala'alate Mountain of north-east strike. In this period, associated structural fractures in basically the same strike with the faults developed.

In Yanshan period, tectonic movement further weakened in intensity till ceased, having little influence on tectonic deformation of Wuxia fault belt (Zhao, 2004). Structural activity of this stage just slightly reformed earlier structures, resulting in formation of a small amount of small-scale shearing fractures induced by faults.

The above four stages of tectonic movement have controlled the occurrence, position and formation and system of structural fractures in Carboniferous-Permian of the Hala'alate Mountain area. The first two stages of tectonic movements happened before and during the formation of the nappe, namely middle-late Hercynian movement. The last two stages of tectonic movements took place after the formation of the nappe, namely Indosinian movement and Yanshan movement. As a result, the structural fractures formed in tectonic movement of earlier stages would change to some extent during the tectonic movement of late stages, while structural fractures formed in the late stages of tectonic movements will remain original state to the maximum extent.

During the four main tectonic stages (especially late Hercynian stage) northwest margin of Junggar Basin experienced, several types of structural fractures caused by compression tectonic movements were formed in formations, among them, regional structural fractures of earlier stages and structural fractures associated with nappe of late stages vertical to original formations are the most important. The regional structural fractures of earlier stages vertical to original formations changed to various degrees in the process of nappe tectonic movement of late stages, and some fractures have even turned nearly horizontal. Therefore, some horizontal fractures or low angle fractures observed in cores nowadays may be regional structural fractures of earlier stages vertical to original formations. The structural fractures associated with nappe of late stages are mainly NE-SW, basically parallel to that of the reverse faults.

The research area, situated in piedmont thrust belt of Wuxia fault belt in the west of Hala'alate Mountain, is rich in nappe structure. From seismic sections interpreted, it can be seen that Carboniferous and Permian encountered in the drilling of the Hala'alate Mountain area are basically foreign nappes, and reverse faults were formed in the same period of the nappes. These reverse faults are different in occurrence and active period, so the structural fractures associated with them are different in occurrence and filling degree too, and have obvious cross-cutting relationship, giving rise to the complicated fracture system in the research area. Based on previous studies on nappes in the Hala'alate Mountain area, actual fracture characteristics in the research area, and the fact that F3 and F4 faults of the latest tectonic movement stage are cut by Triassic at the depression margin, it is concluded that the main forming time of nappes in the Hala'alate Mountain area is Indosinian period, that's to say, Indosinian period is main forming time of structural fractures associated with nappe fault activities in Carboniferous-Permian (especially effective fractures of late stages).

7.3 Fracture effectiveness and distribution

Effective fracture refers to the kind of opening fracture with percolation capacity under natural condition. Once a fracture is filled with other minerals, its percolation capacity will decline, and its effectiveness will decrease correspondingly. Fractures completely filled with minerals can basically be regarded as ineffective fractures; only the fractures not filled with minerals or reopened again after late dissolution and structural transformation can be regarded as effective ones.

Only when the developmental phases of fractures match with the hydrocarbon accumulation stages, can the hydrocarbon migrate from the source rock to volcanic reservoirs through the faults and fractures; besides, only when these fractures still remain open at present, can the hydrocarbon accumulated in the volcanic reservoirs be effectively recovered through the communication of those fractures. Therefore, for the research on hydrocarbon accumulation in Carboniferous and Permian of the Hala'alate Mountain area, it is necessary to define the effective fractures specifically, that is to say the most effective fractures for hydrocarbon accumulation in the research region are the fractures remain open today and have the developmental phases matching with hydrocarbon accumulation stage.

7.3.1 Analysis on fracture effectiveness

Previous researches revealed that with the multi-stage structural evolution, the northwestern margin of Junggar Basin has experienced multi-stage hydrocarbon accumulations (Feng, 2008).

Carboniferous-Permian reservoirs of the Hala'alate Mountain area have experienced mainly 3 hydrocarbon charging stages, including Late Permian-Early Triassic, Late Triassic-Early Jurassic, and Late Jurassic-Early Cretaceous.

The reservoirs' fracture effectiveness analysis is mainly based on the configurationsrelationship of hydrocarbon accumulation periods and fracture formation time.

In Late Permian-Early Triassic, the Permian Fengcheng Formation began to produce hydrocarbon. At the same time, the Hala'alate Mountain area ecperiencedintensive Indosinian tectonic movement. Hydrocarbon from the Permian Fengcheng Formation migrating to shallow ground through active faults. This time is the first hydrocarbon migration and accumulation period of the Hala'alate Mountain area Carboniferous-Permian reservoirs.

In Late Triassic-Early Jurassic, tectonic activities of the northwestern margin of Junggar Basin approached the most active (Feng, 2008).In the meantime, a large number of tectonic fractures were formed in Carboniferous-Permian reservoirs of the Hala'alate Mountain area by the large-scale overthrust activity. Faults in the active period provided good pathways for the hydrocarbon generated by Permian source rocks to migrate to the Hala'alate Mountain area. In this period, Carboniferous-Permian reservoirs in the Hala'alate Mountain area received the second hydrocarbon charging.

In Late Jurassic-Early Cretaceous, some faults in the northwestern margin of Junggar Basin were reactiviated by the second Yanshan tectonic movement. Under the communication of faults, the hydrocarbon generated by Permian source rock charged into the Hala'alate Mountain area again. It is the third important hydrocarbon accumulation stage of Carboniferous-Permian reservoirs in the Hala'alate Mountain area.

Since Late Cretaceous, the third Yanshan tectonic movement has been weakening, northwestern magin of Junngar Basin has entered a stable tectonic evolution stage, and deep-seated faults have stopped activity. Without opening faults connecting source rock, large-scale hydrocarbon accumulation in the Hala'alate Mountain area has been in a near standstill.

The effectiveness of a fracture in Carboniferous-Permian reservoirs of the Hala'alate Mountain area is judged mainly based on its opening degree, oil and gas show degrees, and the relationship between the developmental phase of the fracture and the hydrocarbon accumulation stage.

In terms of the regional geological structurural evolution of northwestern margin of Junngar Basin, the Hala'alate Mountain thrust nappe was formed in Indosinian tectonic movement, thus, the reverse faults in symbiosis with the thrust nappe were formed in Indosinian tectonic movement as well. However, in the process of the formation and evolution of this thrust nappe, the symbiotic reverse faults still have their active ordor. This research shows that the active ordor of these reserves faults is: F2>F6>F3-F5>F4.

Through the structural analysis on the Hala'alate Mountain thrust nappe, it is concluded that Fault F2 slided in plane under horizontal compression and formed the fault-propagation fold in front of the thrust nappe.

As the compressive stress increased, the thrust nappe at the fault-propagation fold broken in plane, giving rise to Fault F6, and associated fractures of NE dip. Created very early, these fractures are high in filling degree, without obvious oil and gas shows, and shown as high resistivity fractures in FMI logging. This kind of fracture was developed in Permian mudstone in this area, which indicates that the NE dipping fractures associated with Fault F6 were created before Late Permian when the source rock of Fengcheng Formation became mature. Because the fractures associated with F6 had already been filled completely before hydrocarbon charging, the formation of these fractures did not match with the first hycrocarbon accumulation stage in this area, so these fractures are considered ineffective.

After the activity of Fault F6 and before the activity of Fault F3, the compression-thrust stress in S-N direction had been increasing during the transition period, net shaped fractures were created in Permian mudstone of the Hala'alate Mountain area, under the intense compression stress. Core observation shows that the net shaped fractures are high in filling degree, almost without oil and gas shows, which indicates that the formation time of the net shaped fractures was close to the fractures associated with F6, and also early than the maturity stage of the source rock of Fengcheng Formation. These net shaped fractures were created and filled very early, and did not match with the main hydrocarbon generation and expulsion stage of source rock in Permain Fengcheng Formation, so these fractures are considered ineffective too.

After the formation of these net shaped fractures, the Hala'alate Mountain thrust nappe continued to evolve, and the fault-propagation fold was dislocated by Fault F3. North-dipping fractures were created with the activity of Fault F3. These fractures in Permian mudstone of this region is characterized by large dip angle and cutting the earlier net shaped fractures of higher filling degree. The fractures of this stage are generally low in filling degree, and filled with a great deal of crude oil and asphalt, which indicates that the formation period of these fractures matches with the main hydrocarbon generation and expulsion stage of the source rock in Permain Fengcheng Formation, thus, it is concluded that the high angle fractures associated with Fault F3 were created during Late Permian-Early Triassic. Combined with the research results of the hydrocarbon inclusions in Carboniferous-Permian reservoirs, it is thought that these fractures might undergo three stages of large scale hydrocarbon accumulation in Late Permian-Early Triassic, Late Triassic-Early Jurassic and Early Jurassic-Late Cretaceous. Because the filling of these fractures was inhibited by the hydrocarbon charging through them, these fractures are still open today, thus, these fractures are considered effective for hydrocarbon accumulation.

To the late evolution stage of the Hala'alate Mountain thrust nappe, Fault F4 and its derived branch faults were active mainly in the shollow strata of the research region, creating unfilled fractures with straight surface in the shallow carboniferous reservoirs in the area. Compared with fractures associated with Fault F3, these fractures have poorer oil and gas shows, and only a small amount of oil stains on the fracture surface (Figure 7-23), so it is inferred that the oil and gas shows were caused by the secondary migration of the early oil in the late adjustment process. Based on the above analysis, the fractures associated with Fault F4 and its derived branch faults were formed after the last stage of hydrocarbon charging. The developmental phase of these fractures did not match with the main hydrocarbon charging stage, and only match with a small quantity of hydrocarbon charging in the reservoir adjustment process in the Hala'alate Mountain area. As a result, these fractures are poorer in effectiveness than the fractures associated with F3, but still effective fractures.

According to hydrocarbon accumulation stages divided according to the inclusions in the fracture fillings metioned in Chapter 5 and fracture effectiveness analysis, it is concluded that fractures associated with Fault F3 happened to provide requisite migration pathways and storage space for hydrocarbon during the 3 main hydrocarbon charging stages, Late Permian-Early Triassic, Late Triassic-Early Jurassic, Late Jurassic-Early Cretaceous, which indicates that the fractures associated with Fault F3 are of great importance for the hydrocarbon accumulation in Carboniferous-Permian reservoirs of the Hala'alate Mountain area. Besides, the fractures associated with Fault F4 and its derived branch faults match with the adjustment period of the Carboniferous-Permian reservoirs in the Hala'alate Mountain area, and have oil and gas shows to some extent, therefore, these fractures are also the key objects in the research on hydrocarbon accumulation in Carboniferous-Permian reservoirs of the Hala'alate Mountain area.

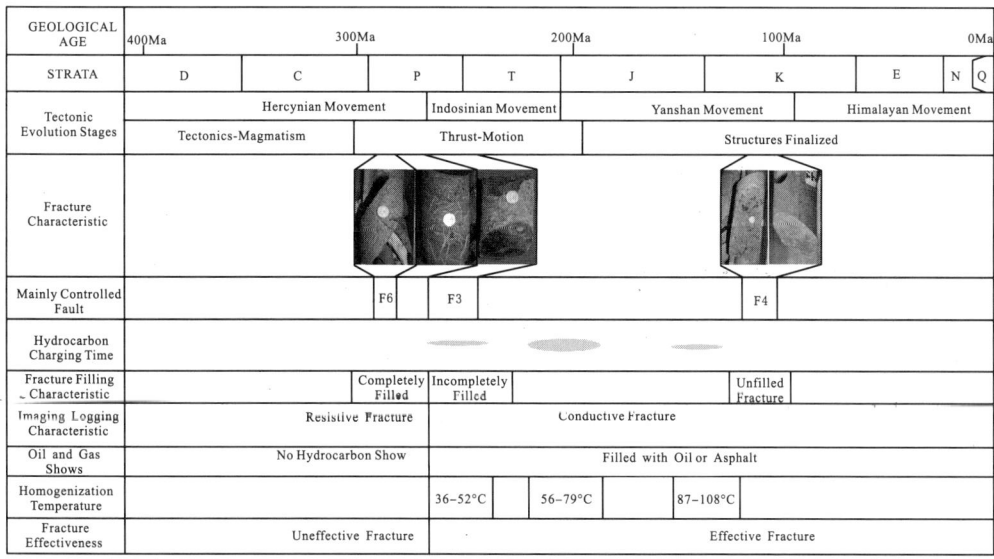

Figure 7-23 Relationship between the developmental phase of the fractures and hydrocarbon accumulation stages in Carboniferous-Permian reservoirs of the Hala'alate Mountain.

7.3.2 Distribution of effective fractures

The Hala'alate Mountain area has experienced several stages of tectonic movements, middle and late Hercynian movement, Indosinian movement, Yanshan movement and so on, in which Indosinian movement had the strongest impact on the formation and distribution of tectonic fractures in Carboniferous-Permian reservoirs of the Hala'alate Mountain area.

On the plane, the maximum horizontal principal stress dictates the development degree of the fractures. The stress orientation is closely related to the orientation of borehole wall collapse and induced fractures. In vertical wells, the maximum or minimum horizontal principal stress can be obtained by analyzing the orientation of borehole wall collapse and induced fractures from the images. In the FMI images, the induced fractures are a group of parallel and 180° symmetrical fracutres with high dip angle, the orientation of which is the maximum horizontal principal stress in present. By picking up the induced fractures from the FMI images of Well HS1, HS2 and HQ6 in the research region, the tectonic stress features around the wells have been analyzed in the research. The statistic results (Table 7-23) show that the maximum horizontal principal stress around Well HS1 and HS2 in the northern research region is NW-SE; while the maximum horizontal principal stress around Well HQ6 in the southeast of the research region is NWW-SEE.

The whole structural configuration of the Hala'alate Mountain area has been controlled by larger regional tectonic movements, and the area has been long affected by Darbut strike-slip fault trending NE, therefore, the main tectonic stress in the research region is NW-SE. This shows that the maximum horizontal principal stress around Well HS1 and HS2 agrees in direction with the main tectonic stress of the research region, but the maximum horizontal principal stress around Well HQ6 is more westerly. The maximum horizontal principal stress dictates the distribution orientation and desity of fractures on the plane. Apparently, the fractures around Well HS1 and HS2 are more close to the main tectonic stress direction of the research region. Besides, these two wells are more close to Darbut strike-slip fault than Well HQ6, as a result, the area where the two wells are experienced more intense tectonic movements, so the fractures are denser there.

Table 7-1 Relationship between the occurrence of drilling-induced fractures and direction of the maximum earth stress in the Hala'alate Mountain area.

Well	Induced fractures			Direction of the maximum earth stress
	Dip angle	Dip	Strike	
HS1	70°–90°	NW	NE–SW	NW–SE
HS2	50°–88°	NW	NE–SW	NW–SE
HQ6	60°–70°	NWW	NNE–SSW	NWW–SEE

It can be seen from the fracture effectiveness analysis of Carboniferous-Permianreservoirs in the research region that the effective fractures closest related to the hydrocarbon accumulation generally have something to do with the activity of Fault F3 and are possibly affected by the activity of Fault F4. That is mainly because that Fault F3, a reverse fault dipping north at high angle, is active over a wide span of formations and its activity period matching with the main hydrocarbon generation period of Permian, so fractures associated with Fault F3 can act as a good conductive system for the vertical migration of hydrocarbon from deep formations to the shallow reservoirs. In contrast, Fault F4 mainly active in shallow formations, can not provide good vertical migration condition for oil and gas, but the unfilled fractures created in the late formation period of Fault F4 can provide space for hydrocarbon accumulation. Besides, it is found from the geochemical analysis of the fracture filling in the research region that the oil source fluid flowed extensively in the research region, especially Well HQ3 and HQ101 in the southern of the research area, the drilling cores from them are abundant in oil and gas shows and the calcite filling fractures in the cores have obvious transformation features by oil source fluid. In contrast, in other wells north of Well HQ3 and HQ101, the calcite filling fractures are created by transformation of two kinds of solution, volcanic hydrothermal solution and atmospheric water, and the drilling cores from them have poorer oil and gas shows. Comprehensive analysis of relationship between developmental phase of fractures and hydrocarbon charging period and fluid geochemical analysis on fracture fillings shows that the effective fractures are mainly distributed in the areas where activity of Fault F3 and F4 overlap on the plane, especially in the area between Well HQ3 and Well HQ101 (Figure 7-24).

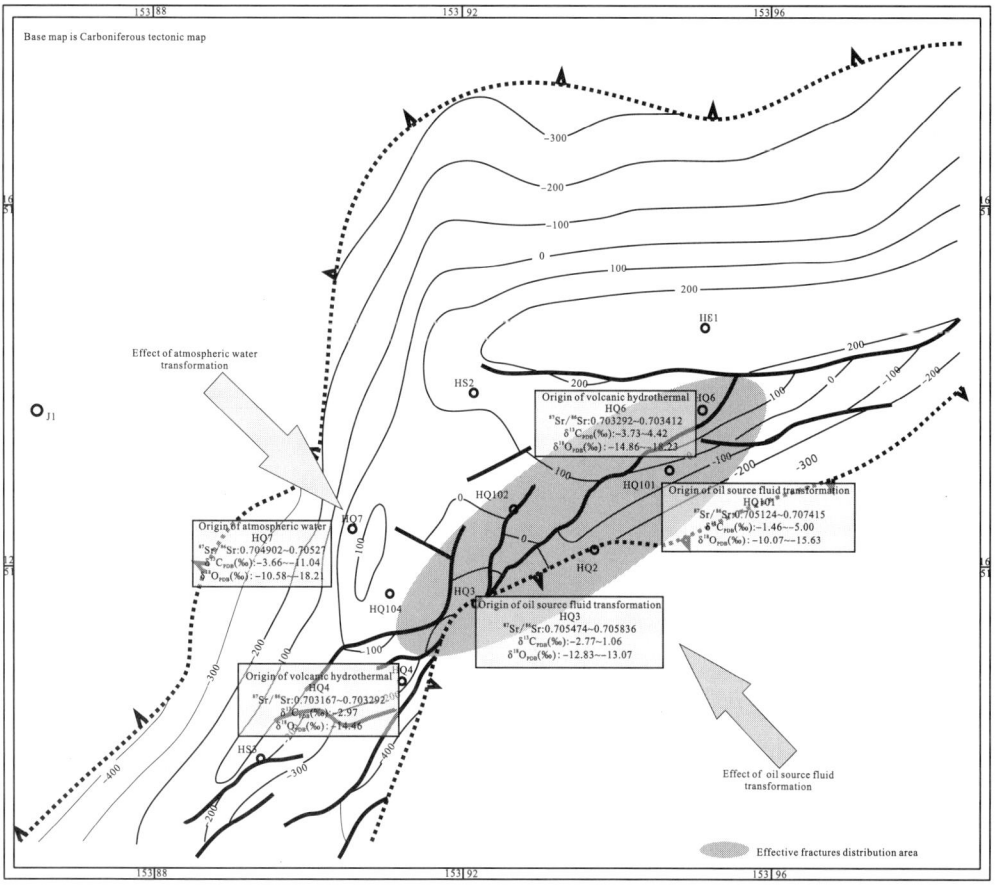

Figure 7-24 Predicted planar distribution of effective fractures in Carboniferous-Permian reservoirs, the Hala'alate Mountain area.

The vertical growth and distribution of the structural fractures can be affected by both formation lithology and fault location in vertical direction. Mechanic characteristic analysis of rock in the research region shows volcanic breccia is the rock most conducive to structural fracture development. Combining the statistic data of the fractures from the regional seismic sections and FMI data reveals that volcanic breccia has higher fracture density than other types of rock in the same tectonic position. However, strong tectonic stress releasing may offset the dominance of lithology over the growth degree of fractures. For example, the volcanic breccia in Well HQ101 has a fracture density of 0.14 m^{-1}, while the mudstone in the fault area with intensive stress releasing has a fracture desity of 0.13 m^{-1} too. Although fracture density is an important parameter in evaluating volcanic reservoirs, it needs to be pointed out fracture effectiveness should be considered more in the evaluation. Therefore, the formations cut through by Fault F3 and F4 vertically are the horizons favorable for the growth of effective fractures, especially intervals of volcanic breccia with higher brittleness.

The analysis above shows the distribution of effective fractures in the research region features: effective fractures are mainly distributed in the areas where Fault F3 and F4 activities overlap on the plane; vertically, mainly in Carboniferous volcanic strata (especially volcanic breccia) Fault F3 and F4 cut through.

References

Banner Jay L. 2004. Radiogenic isotopics: systematics and applications to earth surface processes and chemica l stratigraphy[J]. Earth-Science Reviews, 65 (3-4) : 141-194.

Cai G G, Tong H M. 2010. Analysis on fracture Potential for different types of rocks in Archean Buried Hill: a case study of Liaohe western Sag[J]. Journal of Geomochanics, 16 (3) : 310-320. (in Chinese)

Cao H F, Xia B, Fan L Y, et al. 2007. Formation mechanism and distribution rule of nanyishan fractured reservoirs[J]. Natural Gas Geoscience, 18 (1) : 71-73. (in Chinese)

Cao J, Hu W X, Yao S P, et al. 2007. Carbon, oxygen and strontium isotope composition of calcite veins in the carboniferous to perm ian source sequences of the Junggar Basin: implications on petroleum fluid migration[J]. Acta Sedimentologica Sinica, 25 (5) : 722-729. (in Chinese)

Chen F J, Wang X W, Wang X W. 2005. Prototype and tectonic evolution of the Junggar Basin, northwestern China[J]. Earth Science Frontiers, 12 (3) : 77-89. (in Chinese)

Chen X F, Kuang J. 2010. Distribution of carboniferous source rocks and petroleum systems in the Junggar Basin[J]. Petroleum Exploration And Development, 37 (4) : 397-408.

Chen Y Q, Wang W F. 2004. Structural evolution and pool-forming in Junggar Basin[J]. Journal of the University of Petroleum, China (Editon of Natural Science), 28 (3) : 4-9. (in Chinese)

Chen Z H, Zha M, Jin Q, et al. 2011. Distribution and characteristics of the homohopane molecular parameters in paleogene system of the Dongying Sag[J]. Acta Sedimentologica Sinica, 29 (1) : 173-183. (in Chinese)

Clark I D, Fritz P. 2000. Environmental Isotopes ın Hydrogeology[M]. New York: Lewis Publishing House.

Craig H. 1966. Isotope composition and origin of the Red Sea and Slton Sea geo thermal brines[J]. Science, 154 (3756) : 1544-1548.

Criss R E, Gregory R T, Taylor H P. 1987. Kinetic theory of oxygen isotope exchange between minerals and water[J]. Geochim Cosmochim Acta, 51 (5) : 1099-1108.

Dai J S, Xu J C, Meng Z P et al. 2003. Prediction of volcanic rock fissure with finite deformation method[J]. Journal of The University of Petroluem, China (Edition of Natural Science), 27 (1) : 1-3+10-9. (in Chinese).

Dai Y Q, Luo J L, Lin T, et al. 2007. Reservoir characteristics and petrogenesis of volcanic rocks in the Yingcheng formation of the Shengping gas field, northern Songliao basin[J]. Geology in China, 34 (3) : 528-535. (in Chinese)

Deines P. 1980. The isotopic composition of reduced organic carbon[A]//Fritz P, Fontes J C. Handbook of Environmental Isotope geochemistry, 1. The Terrestrial Environment[M]. Amsterdam Elsevier, 329-345.

Ding A N, Hui R Y, Wang Y T. 1994. Organic petrography of hydrocarbon source rock of the carbonniferous and permian in northwestern margin, Junggar Basin [J]. Xing Jiang Petroluem Geology, 15 (3) : 220-225. (in Chinese)

Ding W L, Li C, Li C Y, et al. 2012. Dominant factor of fracture development in shale and its relationship to gas accumulation[J]. Earth Science Frontiers, 19 (2) : 212-220. (in Chinese)

Dorbon M, Schmitter J M, Garrigues P, et al. 1984. Distribution of carbazole derivatives in petroleum[J]. Organic Geochemistry, 7(2): 111-120.

Du X B, Zhang C, Zhang C M. 2005. Present application situation and prospect of stable isotope geochemistry during basin fluid analysis[J]. Petroleum Geology and Recovery Efficiency, 12(4): 20-22+82-83. (in Chinese)

Emery D, Robinson A. 1993. Inorganic geochemistry: application to petroleum geology[J]. London: Blackwell Scientific Publications.

Fan C H, Qin Q R, Zhi D M, et al. 2012. Controlling factors and characteristic of carboniferous volcanic reservoir fractures of Zhongguai Uplift in northwestern margin of Junggar Basin[J]. Natural Gas Geoscience, 23(1): 81-87. (in Chinese)

Fan C, Su Z, Zhou L. 2014. Kinematic features of darlbute fault in northwestern margin of Junggar Basin[J]. Chinese Journal of Geology, 49(4): 1045-1058. (in Chinese)

Feng J W. 2008. the tectonic evolution and it's controlling on hydrocarbon in Wuxia Fault Belt of Junggar Basin[D]. Beijing: China University of Petroleum (Beijing). (in Chinese)

Gao C H, Zha Ming, Qu Jiangxiu, et al. 2015. Fluid inclusion characteristics and hydrocarbon accumulation stages of unconformable reservoirs in the northwest margin of the Junggar Basin[J]. Natural Gas Industry, 35(11): 23-32. (in Chinese)

Gao Q D, Zhao K Z, Hu X F, et al. 2011. C-O, Sr isotope composition of the carbonate in Ordovician in Tarim Basin and implication on fluid origin[J]. Journal of Zhejiang University(Science Edition), 38(5): 579-583. (in Chinese)

Gao X Z, Chen F J. 2000. Application of fluid inclusions to determination of the times and stages of hydrocarbon reservoir filling: a case study of Nanbaxian oilfield in the Qaidam Basin[J]. Earth Science Frontiers, 7(4): 548-554. (in Chinese)

Gao X, Xie Q B. 2007. Advances in identification and evaluation of fracture[J]. Progress in Geophysics, 22(5): 1460-1465. (in Chinese)

Han B F, Ji J Q, Song B, et al. 2006. Late Paleozoic vertical growth of continental crust around the Junggar Basin. Xinjiang. China: Timing of post-collisional plutonism[J]. Acta Petrologica Sinica, 22(5): 1077-1085. (in Chinese)

Han L G, Zhang Z H, Li W. 2006. An analysis of the present oil migration direction of block I in central Junggar Basin[J]. Acta Geoscientica Sinica, 27(4): 335-340. (in Chinese)

Handin J, Hager R V J. 1958. Experimental deformation of sedimentary rocks under confining pressures: tests at high temperature[J]. AAPG Bulletin, 42(12): 2892-2934.

Hao F, Zhou X H, Zhu Y M, et al. 2009. Mechanisms for oil depletion and enrichment on the Shijiutuo uplift, Bohai Bay basin, China[J]. AAPG Bulletin, 93(8): 1015-1037.

He D F, Chen X F, Kuang J, et al. 2010. Characteristics and exploration potential of Carboniferous hydrocarbon plays in Junggar Basin[J]. Acta Petrolei Sinica, 31(1): 1-11.

He D F, Guan S W, Zhang N F, et al. 2006. Thrust belt structure and significance for petroleum exploration in Hala' alat Mountain in northwestern margin of Junggar Basin[J]. XinJiang Petroleum Geology, 27(3): 267-269. (in Chinese)

He D F, Zhai G M, Kuang J. 2005. Distribution and Tectonic Tectonic Features of Paleo-Uplifts in the Junggar Basin[J]. Chinese Journal Of Geology(Scientia Geologica Sinica), 40(2): 248-261. (in Chinese)

Hoefs J. 1997. Stable Isotope Geochemistry (4th Edition) [M]. Berlin: Springer-Verlag.

Hu Y, Xia B. 2012. An approach to the tectonic evolution and hydrocarbon accumulation in the Halaalate Mountain area, northern Xinjiang[J]. Sedimentary Geology and Tethyan Geology, 37(2): 52-58. (in Chinese)

Hu Z W, Huang S J, Wang C M, et al. 2009. Application of strontium isotope geochemistry to the oil and gas reservoir diagenesis research[J]. Contributions to Geology and Mineral Resources Research, 24(2): 160-165. (in Chinese)

Huang J S, Shi H, Zhang M, et al. 2002. Application of strontium isotope stratigraphy to diagenesis research[J]. Acta Sedimentologica Sinica, 20(3): 359-366. (in Chinese)

Jiang Z S, Fowler M G. 1986. Carotenoid-derived alkanes in oils from northwestern China[J]. Organic Geochemistry, 10(4-6): 831-839.

Jrgen J. 1988. Carbon and oxygen isotopic studies of the chalk reservoir in the Skjold oilfield, Danish North Sea: implications for diagenesis[J]. Chemical Geology, 73(2): 97-107.

Ju W, Hou G T, Feng S B, et al. 2014. Quantitative prediction of the Yanchang formation Chang 63 reservoir tectonic in the Qingcheng-Heshui Area, Ordos Basin. Earth Science hrontiers, 21(6): 310-320. (in Chinese)

Koch P L, Zachos J C, Gingerich P D. 1992. Correlation between isotope records in marine and continental carbon reservoirs near the Palaeocene boundary[J]. Nature, 358(6384): 319-322.

Lai S H, Yu Q, Zhou W, et al. 2004. Development period of fractures in the Late Triassic-Jurassic in the north Chuxiong Basin[J]. Petroleum Exploration and Development, 31(5): 25-29. (in Chinese)

Lei Z Y, Lu B, Wei Y J, et al. 2005. Tectonic evolution and development and distribution of fans on northwestern edge of Junggar Basin[J]: Oil & Gas Geology, 26(1): 86-91. (in Chinese)

Li G L. 2013. Analysis of carboniferous igneous hydrocarbon accumulation characteristics and controlling factors in the Hala' alate Mountain area[D]. Qingdao: China University of Petroleum(Huadong). (in Chinese)

Li J Y, Xiao X C, Chen W. 2000. Late Ordovician continental basement of the eastern Junggar Basin in Xinjiang. NW China: Evidence from the Laojunmiao metamorphic complex on the northeast basin margin[J]. Regional Geology of China, 19(3): 297-302. (in Chinese)

Li J Y, Xiao X C. 1999. Brief review on some issues of Framework and Tectonic evolution of XinJiang Crust, NW China[J]. Scientia Geologica Sinica, 34(4): 405-419. (in Chinese)

Li J, Xu P H, ANG A Q, Liu X Y. 2008. Characteristics and controlling factors of Carboniferous volcanic reservoir in the middle section of the northwestern margin of Junggar Basin[J]. Acta Petrolei Sinica, 29(3): 329-335. (in Chinese).

Li M, Larter S R, Stoddart D, et al. 1995. Fractionation of pyrrolic nitrogen compounds in petroleum during migration: derivation of migration-related geochemical parameters[J]. Geological Society, London, Special Publications, 86(1): 103-123.

Li W, Hu J M, Qu H J. 2009. Discussion on Mesozoic basin boundary of the northern Junggar Basin, Xinjiang[J]. Journal of Northwest University(Natural Science Edition), 39(5): 821-830. (in Chinese)

Li Y H. 1998. Some apllications of Isotope-tracing in geology[J]. Earth Science Frontiers, 5(2): 106-112. (in Chinese)

Lin Z B, Wu X H, Wang Y, et al. 2006. The structural belt compartment of carboniferous basement and petroleum distribution in Junggar Basin[J]. Xinjiang Petroleum Geology, 27(4): 389-393. (in Chinese)

Liu C G, Li G R, Zhang Y W, et al. 2007. Application of strontium isotope to the study of paleokarst——an case from ordovician in the Tahe Oilfield, Tarim Basin[J]. Acta Geologica Sinica, 81(10): 1398-1406. (in Chinese)

Liu C L. 1998. Carbon and oxygen isotopic compositions of lacustrine carbonates of the Shahejie Formation in the Dongying depression and their paleolimnological significance[J]. Acta Sedimentologica Sinica, 16(3): 109-114. (in Chinese)

Liu C Z, Sun Y K, Yu H S, et al. 2010. Study on characteristic of carboniferous volcanic oil and gas reservoirs and alkaline diagenesis in the Santanghu Basin, NW China[J]. Journal of Jilin University(Earth Science Edition), (6): 1221-1231. (in Chinese)

Liu D G, Xiao X M, Tian H, et al. 2008. Fluid inclusion types and their geological significance in Petroliferous basins[J]. Oil & Gasgeology, 29(4): 491-501. (in Chinese)

Liu J Y, Xi A H, Ran Q Q et al. 2012. Petrography characteristics and main controlling factors of secondary pores in Carboniferous volcanic reservoir in Dixi area, Junggar Basin[J]. Northwest Oil & Gas Exploration, 24(3):51-55. (in Chinese)

Liu L, Sun X M, Dong F X, et al. 2004. Geochemical characteristics and fluid inclusion in calcite viens of lower part of member 1 of Shahejie Formation, off shore area, Dagang oilfield: a case study of well Gangshen 67[J]. Journal of Jilin University (Earth Science Edition), 34(1):49-54. (in Chinese)

Liu Z. 2012. Formation and evolution and its structural modeling of Hala'alate thrust nappe at the northwestern margin of Junggar Basin[D]. Beijing: China University of Geosciences (Beijing), 1-83. (in Chinese)

Lu B, Zhang J, Li T, et al. 2008. Analysis of tectonic framework in Junggar Basin[J]. Xinjiang Petroleum Geology, 28(5):283-289. (in Chinese)

Lu J G, Wu Z L, Guan X C. et al. 2004. Automatically extract fracture parameters from resistivity images by using hough transform[J]. Well Logging Technology, 28(2):115-117+179. (in Chinese)

Meng Z P, Peng S P, Ling B C. 2000. Character of the deformation and Strength under different confining pressure on sedimentary rock[J]. Journal of China Coal, 1(25):15-18. (in Chinese)

Morrow D. 1982. Diagenesis: dolomite part I, the chemistry of dolomitization and dolomite precipitation[J]. Geoscience Canada, 1(9):5-13.

Ohm S E, Karlsen D A, Austin T J F. 2008. Geochemically driven exploration models in uplifted area: examples from the Norwegian Barents Sea[J]. AAPG Bulletin, 92(9):1191-1223.

Othman R, Ward C R, ArouriK R. Oil generation by igneous intrusions in the northern Gunnedah basin, Australia[J]. Organic Geochemistry, 32(10):1219-1232.

Palmer M R, Edmond J M. 1989. The strontium isotope budget of the modern ocean[J]. Earth Planet Science Letter. 92(1):11-26.

Palmer M R. Elderfield H. 1985. Sr isotope composition of seawater over the past 75 Ma[J]. Nature, 314(6011):526-528.

Peters K E, Moldowan J M. 1993. The Biomarker Guide: Interpreting Molecular Fossils in Petroleum and Ancient Sediments[M]. New Jersey: Prentice Hall, 363-365.

Peters K E. Walters C C, Moldowan J M. 2005. The Biomarker Guide (2nd, Volume 2): Biomarkers and Isotopes in Petroleum Exploration and Earth History[M]. Cambridge: Cambridge University Press, 2-15.

Qin Q R, Su P D, Wu M J, et al. 2008. Fracture identication in igneous rock reservoirs of the ninth block at the northwest margin of the Junggar Basin[J]. Natural Gas Industry, 28(5):25-29. (in Chinese)

Qu Y. 2015. The pore structures and Percolation Characteristics of Intermediate-basic Volcanic Reservoirs in Xushen Gas Field[J]. Journal of Yangtze University (Natural Science Edition), 12(20):7-11+3. (in Chinese)

Rohrman M. 2003. Hydrocarbon potential of volcanic basins: principles and rules of thumb, hydrocarbons in crystalline rocks[J]. Geol Soc Lon Spec Pub, 214:7-33.

Ruan B T, Zhang J H, Wang Z W et al. 2011. The affected factor for volcanic fracture development[J]. Natural Gas Geoscience, 22(2):287-292. (in Chinese)

Rushdy O, Khaled R A, Colin R. 2002. Oil generation by igneous intrusions in the northern Gunnedah basin, Australia[J]. Organic Chemistry, 32(10):1218-1230.

Schutter S R. 2003. Occurrences of hydrocarbons in and around igneous rocks[J]. Geo Soc Lon Spec Pub, 214(1):35-68.

Shi J, Zou N N, Lu X C, et al. 2013. Geochemical characteristics and genetic mechanism of permian dolomitic clastic rocks in northwestern Junggar basin[J]. Acta Sedimentologica Sinica. 31(5):898-906. (in Chinese)

Suchy Y, Heijlen W, Sykorova I. 2000. Geochemical study of calciteveins in the silurian and devonian of the Barrandian Basin (Czech Republic): evidence for widesp read post-Variscan fluid flow in the central part of the Bohemian Massif[J]. Sedimentary Geology, 13(3-4): 201-219.

Sui F G. 2013. The Practices for exploration breakthrough in Sinopec Prospecting Area in northwestern margin of Junggar Basin[J]. Xinjiang Petroleum Geology, 34(2): 129-132. (in Chinese)

Tan K J, Zhang F, Wu X Z, et al. 2008. Basin-range coupling and hydrocarbon accumulation at the northwestern margin of the Junggar Basin[J]. Natural Gas Industry, 28(5): 10-13. (in Chinese)

Taylor H P J, Frechen J, Degens E T. 1976. Oxygen and carbon isotope studies of carbonatites from the Laacher See district, West Germany and the Alno district, Sweden[J]. Geochimica et Cosmochimica Acta, 31(3): 407-430.

Tian J Q, Zou H Y, Xu C G, et al. 2011. Application of ETR in oil-source correlation for severely biodegradaed crude oil—by taking JX1-1 oilfield for example[J]. Journal of Oil and Gas Technology, 33(7): 19-23+36. (in Chinese)

Toyoda K, Horiuchi H, TokonamiM. 1994. Dupal anomaly of Brazilian carbonatites: geochemical correlationswith hotspots in the South Atlantic and imp lications for the mantle source[J]. Earth and Planetary Science Letter, 126(4): 315-331.

Veizer J, Buhl D, Diener A. 1997. Strontium isotope stratigraphy: potential resolution and event correlation[J]. Palaeogeography, Palaeoclimatology, Palaeoecology, 132(1): 65-77.

Veizer J, Hoefs J. 1976. The nature of $^{18}O/^{16}O$ and $^{13}C/^{12}C$ secular trends in sedimentary carbonate rocks[J]. Geochimica et Cosmochimica Acta, (40): 1387-1395.

Wang D R, Zhang Y H. 2001. A study on the origin of the carbonate cements within reservoirs in the external metamorphic belt of the Bohai Bay oil-gas bearing region[J]. Petroleum Exploration and Cevelopment, 28(2): 40-42+109-110+118-119. (in Chinese)

Wang G L, Wan F G, Zhua A G. 1989. Resources evaluatiom by mathematical simulation in the region of Mafu Depression, Junggar Basin[J]. Xinjiang Petroleum Geology, 10(3): 100-112. (in Chinese)

Wang G Q, Yue Y F, Qi J F, et al. 2002. A fracture analysis of paleogene deep lastic reservoir rock in baitangma area, Huanghua depression[J]. China Offshore Oil and Gas(Geology), 16(6): 21-25. (in Chinese)

Wang G Z, Liu S G. 2009. Paleo-fluid geochemical evaluation of hydrocarbon preservation in marine carbonate rock areas: taking lower association in central Sichuan Basin as an example[J]. Journal of Chengdu University of Technology (Science & Technology Edition), 36(6): 631-644. (in Chinese)

Wang R C, Xu H M, Shao Y et al. 2008. Reservoir characteristics of Carboniferous volcanic rocks in Ludong area of Junggar Basin[J]. Acta Pertolei Sinica, 29(3): 350-355. (in Chinese)

Wang S Z, Zhang K H, Jin Q. 2014. The genntic types of crude oils and the petroleum geological significance of the Fengcheng formation source rock in the Hala'alate Mountain area, Junggar Basin[J]. Natural Gas Geoscience, 25(4): 595-602. (in Chinese)

Wang X L, Tang Y, Chen Z H, et al. 2013. Carboniferous Lithofacies Paleogeography in the North of Xinjiang[J]. Acta Sedimentologica Sinica, 31(4): 571-579. (in Chinese)

Wu K Y. 2009. Research on the stages of reservoir formation in Wuerhe-Xiazijie Area in Junggar Basin[J]. Journal of Oil and Gas Technology, 3(31): 18-23. (in Chinese)

Xiong Y X, Ai H, Ran Q Q, et al. 2012. The formation mechanism and four-stage evolution of volcanic primary reservoir spaces: a case study of Carboniferous volcanic rocks in Di xi area, Junggar basin[J]. Geology in China, 39(1): 146-155. (in Chinese)

Xu X Y, Li R S, Chen J L, et al. 2014. New constrains on the paleozoic tectonic evolution of the northern Xinjiang area[J]. Acta Petrologica Sinica, 30(6): 1521-1534. (in Chinese)

Yang G, Wang X B, Li B L, et al. 2011. Transpression and wrench faults of northwestern margin of Junggar Basin[J]. Chinese Journal of Geology, 46(3):696-708. (in Chinese)

Yang H B, Chen L, Kong Y H. 2004. A novel classification of structural units in Junggar Basin[J]. Xinjiang Petroleum Geology. 26(6):686-688. (in Chinese).

Yuan H F, Liu Y, Xu F H, et al. 2014. The fluid charge and hydrocarbon accumulation, Sinian reservoir, Anpingdian-Gaoshiti Structure, Central Sichuan Basin[J]. Acta Petrologica Sinica, 30(3):727-736. (in Chinese)

Zhang K H, Lin H X, Zhang G L et al. 2015. Characteristics and controlling factors of volcanic reservoirs of Hala' alate mountains tectonic belt[J]. Journal of China University of Petroleum (Edition of Natural Science), 39(2):16-22. (in Chinese)

Zhang S W. 2013. Identification and its petroleum geologic significance of the Fengcheng formation source rocks in Hala' alt area, the northern margin of Junggar Basin[J]. Oil & Gas Geology, 34(2):145-152. (in Chinese)

Zhang X L. 1985. Relationship between carbon and oxygen stable isotope in carbonate rocks and paleosalinity and paleotemperature of seawater[J]. Acta Sedimentologica Sinica, 3(4):17-30. (in Chinese)

Zhang Y J, Cao J, Hu W X. 2010. Timing of petroleum accumulation and the division of reservoir-forming assemblages, Junggar Basin, NW China[J]. Petroleum Exploration and Development, 37(3):257-262. (in Chinese)

Zhao B. 2004. Effect of Yanshan movement and himalayan orogeny on hydrocarbon migration and accumulation in Junggar Basin[J]. Xinjiang Petroleum Geology, 25(5):468-470. (in Chinese)

Zhao H L, Liu Z W, Li J et al. 2004. Petrologic characteristics of igneous rock reservoirs and theirresearch orientation[J]. Oil&Gas Geology, (6):609-613. (in Chinese)

Zhao N and Shi Q. 2012. Characteristics of fractured and porous volcanic reservoirs and the major controlling factors of their physical properties: a case studyssss from the carboniferous volcanic rocks in Ludong-Wucaiwan area. Junggar Basin[J]. Natural Gas Industry, 32(10):14-23+108-109. (in Chinese)

Zhao Y C, Zhou X F, Wang C X, et al. 2005. Characters of a Special rock-fractured reservoir and factors of controlling fractured development at Qingxi oil field in Jiuxi Basin[J]. Natural Gas Geosicence, 16(1):12-15. (in Chinese)

Zheng L B, Gong L, Zu K W, Tang X M, et al. 2012. Influence factors on fracture validity of the paleogene Reservoir, Western Qaidam Basin[J]. Acta Geologica Sinica, 86(11):1809-1814. (in Chinese)

Zheng Y H, Huang H P, Wen Z G, et al. 2004. Oil Migration direction of daluhu oilfield in dongying depression[J]. Natural Gas Geoscience, 15(6):650-651. (in Chinese)

Zhou X G, Cao C J, Yuan J Y. 2003. The research actuality and major progresses on the quantitative forecast of reservoir fractures and hydrocarbon migration law[J]. Advance In Earth Science, 18(3):398-404. (in Chinese)

Zhu D Y, Zhang D W, Zhang R Q, et al. 2015. Fluid alteration mechanism of dolomite reservoirs in Dengying Formation, South China[J]. Acta Petrolei Sinica, 36(10):1188-1198. (in Chinese)

Zhu S F, Liu X, Zhu X M, et al. 2015. The formation mechanism of reservoir differences between the Hanging Wall and the foot wall of Ke-Bai overthrust Fault, Junggar Basin[J]. Acta Sedimentologica Sinica, 33(1):194-201.

Zou C N, Zhao W Z, JIA C Z et al. 2008. Formation and distribution of Volcanic hydrocarbon reservoirs in sedimentary basins of China[J]. Petroluem Explorationg and Development, 25(3):257-271. (in Chinese)